INTRODUCING FORTRAN 90

Springer
Berlin
Heidelberg
New York
Barcelona
Budapest
Hong Kong
London
Milan
Paris
Tokyo

Ian Chivers and Jane Sleightholme

INTRODUCING
FORTRAN 90

 Springer

Ian David Chivers, B.Sc., M.Sc., M.B.C.S. C.Eng.
Jane Sleightholme, M.Sc.

The Computer Centre
King's College, Strand
London WC2R 2LS, UK

ISBN 3-540-19940-3 Springer-Verlag Berlin Heidelberg New York

British Library Cataloguing in Publication Data
Chivers, I. D.
   Introducing Fortran 90
   I. Title II. Sleightholme, Jane
   005.133

   ISBN 3-540-19940-3

Library of Congress Cataloging-in-Publication Data
A catalog record for this book is available from the Library of Congress

Apart from any fair dealing for the purposes of research or private study, or criticism or review, as permitted under the Copyright, Designs and Patents Act 1988, this publication may only be reproduced, stored or transmitted, in any form or by any means, with the prior permission in writing of the publishers, or in the case of reprographic reproduction in accordance with the terms of licences issued by the Copyright Licensing Agency. Enquiries concerning reproduction outside those terms should be sent to the publishers.

© Springer-Verlag London Limited 1995
Printed in Great Britain

The use of registered names, trademarks, etc. in this publication does not imply, even in the absence of a specific statement, that such names are exempt from the relevant laws and regulations and therefore free for general use.

The publisher makes no representation, express or implied, with regard to the accuracy of the information contained in this book and cannot accept any legal responsibility or liability for any errors or omissions that may be made.

Typesetting: Camera ready by authors
Printed and bound at the Athenæum Press Ltd., Gateshead, Tyne and Wear
34/3830-543210 Printed on acid-free paper

to Joan, Martin and Jenny

to Steve, Mark, and Jonathan

Preface

This book has evolved from our combined experience of working in computing services at the University of London (for the last nine years at King's College, and before that eight years at Imperial College and seven at Chelsea College) in the teaching, advice and technical support of Fortran and related areas.

Thanks are due to:–

- the staff and students at King's College London – without them none of this would have been possible; also the support and facilities provided by the Computer Centre;

- the patience of our families during the lengthy period required to develop the courses upon which this book is based and whilst preparing the camera ready copy;

- the staff at NAG, Salford Fortran and DEC for their support. Special thanks to Steve Lionel at DEC and Tim Bartle at Salford for the opportunity to take part in the beta testing of the Alpha compiler and the Salford Nag compiler respectively. The lessons to be learnt from moving programs between the three compilers were invaluable;

- the people on comp.lang.fortran and the specialist Fortran 90 list. Access to the expertise of several hundred people involved in the use of Fortran on a daily basis across a wide range of disciplines is inestimable;

- the staff at the BSI, IEC and ISO – Annex D from ISO/IEC 1539:1991 has been reproduced with the permission of the International Organization for Standardization, ISO and the International Electrotechnical Commission, IEC. The complete standard can be obtained from any ISO or IEC member or from the ISO or IEC Central offices, Case Postal, 1211 Geneva 20, Switzerland. Copyright remains with ISO and IEC.

- finally Beverley Ford at Springer Verlag for her infectious enthusiasm and encouragement during the gestation period and birth!

The course notes and book were prepared on a range of PCs using a variety of versions of Corel Ventura Publisher. The facilities provided by Ventura are superb, and make the production of firstly the notes and secondly the book straightforward. Whilst Ventura has a slightly steeper learning curve than that of word processing software the investment is quickly repaid in the time and effort saved. Draft output was to a range of Postscript laser printers. Final output was to a Linotron 300 typesetter at the University of London Computer Centre.

Preface

Tools that proved invaluable were:–

- EC. This is a programmers screen editor. It is small and extremely fast and provided a very good tool for the development of Fortran programs. Edlin and EDIT are not ideal for this!

- Xtree Gold. This is a very useful file file management utility for DOS and the PC. Keeping track of approaching a thousand files over two years on a number of PCs is not without problems. Xtree helped out tremendously here.

- The Mortice Kern Systems Unix Toolkit. This provides UNIX under DOS and Windows on a PC. The pattern matching facilities provided by ED and SED provide a very powerful set of tools for the manipulation of text files. Other tools that proved invaluable include HEAD, TAIL, AWK, SORT, SPELL and the Korn Shell.

- A range of networking software that made the transfer of the programs and data files between the various hardware platforms easy. Most notably the FTP and Telnet facilities provided by WINQVT over IP, but also Kermit in the days of X 25 connectivity.

A range of hardware platforms, operating systems and Fortran compilers were used. These were:–

- DEC VAX under VMS and later OPEN VMS with the NAG Fortran 90 compiler;

- IBM PC and compatibles under DOS and Windows using the Salford NAG Fortran 90 compiler;

- DEC Alpha under OPEN VMS using the DEC Fortran 90 compiler;

We are currently in the process of putting all program examples up at King's on our WWW server. The material will be found under:–

http://www.kcl.ac.uk/kis/support/cc/courses

but the exact location is not finalised at this time. This material will be made available as simple ASCII text, to minimise compilation problems. New course material is under development and will be put up as and when it is available.

Ian D Chivers, i.chivers@kcl.ac.uk

Jane Sleightholme, j.sleightholme@kcl.ac.uk

1995.

Contents

1	**Overview**	**1**
2	**Introduction to Computer Systems**	**6**
2.1	The components of a computer system	7
2.2	Software	8
2.3	Problems	9
2.4	Bibliography	9
3	**Introduction to Operating Systems**	**11**
3.1	History of Operating Systems	12
3.1.1	The 1940s	12
3.1.2	The 1950s	12
3.1.3	The 1960s	12
3.1.4	The 1960s and 1970s	12
3.1.5	The 1970s and 1980s	13
3.2	Networking	13
3.3	Problems	14
3.4	Bibliography	14
4	**Introduction to Using a Computer System**	**15**
4.1	Files	16
4.2	Editors	16
4.3	Stand Alone Systems	16
4.4	Networked Systems	16
4.5	Multi-User Systems	17
4.6	Other Useful Things to Know	17
4.7	Bibliography	18
5	**Introduction to Problem Solving**	**19**
5.1	Natural Language	20
5.2	Artificial Language	21
5.2.1	Notations	21
5.3	Resume	21
5.4	Algorithms	22
5.4.1	Top Down	22
5.4.2	Bottom up	22
5.4.3	Stepwise Refinement	23
5.5	Systems Analysis and Design	23
5.5.1	Problem Definition	24
5.5.2	Feasibility Study and Fact Finding	24
5.5.3	Analysis	24
5.5.4	Design	24
5.5.5	Detailed Design	24

5.5.6		Implementation	25
5.5.7		Evaluation and testing	25
5.5.8		Maintenance	25
5.6	Conclusions		25
5.7	Problems		26
5.8	Bibliography		26
6	**Introduction to Programming Languages**		**28**
6.1	Some Early Theoretical Work		29
6.2	What is a programming language ?		29
6.3	Program Language Development and Engineering		29
6.4	The Early Days		29
6.4.1		Fortran	30
6.4.2		Cobol	30
6.4.3		Algol	31
6.5	Chomsky and Program Language Development		31
6.6	Lisp		32
6.7	Snobol		32
6.8	Second Generation Languages		32
6.8.1		PL/1 and Algol 68	32
6.8.2		Simula	33
6.8.3		Pascal	33
6.8.4		APL	33
6.8.5		Basic	34
6.8.6		C	34
6.9	Some Other Strands in Language Development		34
6.9.1		Abstraction, Stepwise Refinement and Modules	34
6.9.2		Structured Programming	34
6.9.3		Standardisation	35
6.10	Ada		35
6.11	Modula		35
6.12	Modula 2		36
6.13	Other Language Developments		36
6.13.1		Logo	36
6.13.2		Postscript, TeX and LaTeX	37
6.13.3		Prolog	37
6.13.4		SQL	37
6.13.5		ICON	37
6.14	Object Orientated Programming – OOP		38
6.14.1		Oberon and Oberon 2	38
6.14.2		Smalltalk	39

6.14.3	C++	39
6.15	Fortran 90	40
6.16	The Future and Further Sources	40
6.16.1	Fortran 1996	40
6.16.2	High Performance Fortran – HPF	41
6.16.3	Network Sources	41
6.17	Summary	42
6.18	Bibliography	42

7 Introduction to Programming — 47

7.1	Elements of a programming language	49
7.2	Variables — name, type and value	52
7.3	Notes	54
7.4	Some more Fortran rules	55
7.5	Good Programming Guidelines	56
7.6	Fortran Character set	56
7.6.1	Notes	56
7.7	Problems	56

8 Introduction to Arithmetic — 58

8.1	Rounding and truncation	61
8.2	Example 1: Time taken for light to travel from the Sun to Earth	63
8.3	The PARAMETER statement.	64
8.4	Precision and size of numbers	65
8.5	Health Warning: Optional Reading, Beginners are Advised to Leave until Later	67
8.5.1	Selecting different INTEGER Kinds	69
8.5.2	Selecting different REAL Kinds	69
8.5.3	Specifying Kind Types for Literal Integer and Real Constants	70
8.5.4	Positional Number Systems	70
8.5.5	Bit Data Type and Representation Model.	71
8.5.6	Integer Data Type and Representation Model	71
8.5.7	Real Data Type and Representation Model	72
8.5.8	IEEE 754	72
8.5.9	Example 2: Testing the numerical representation of different kind types on a system	72
8.5.10	Example 3: Binary Representation of Different Integer Kind Type Numbers	75
8.5.11	Summary of how to select the appropriate KIND type	77
8.6	Summary	77
8.7	Problems	78
8.8	Bibliography	80

9 Arrays 1: Some Fundamentals — 81

9.1	Tables of data	82

9.1.1	Telephone directory	82
9.1.2	Book catalogue	82
9.1.3	Examination marks or results	83
9.1.4	Monthly rainfall	83
9.2	Arrays in Fortran	84
9.3	The DIMENSION Attribute	84
9.4	An index	84
9.5	Control structure	85
9.6	Monthly Rainfall	85
9.6.1	Example 1: Rainfall	86
9.7	People's Weights	87
9.7.1	Example 2: People's Weights	88
9.8	Summary	88
9.9	Problems	89

10 Arrays 2: Further Examples — 90

10.1	Higher dimension arrays	91
10.1.1	Example 1: A Map	91
10.1.2	Example 2: Booking arrangements in a theatre or cinema	92
10.2	Additional forms of the DIMENSION attribute and DO loop statement	93
10.2.1	Example 3: Voltage from -20 to +20 volts	93
10.2.2	Example 4: Longitude from -180 to +180	94
10.2.3	Notes	94
10.3	The DO loop and straight repetition	94
10.3.1	Example 5: Table of Temperatures	94
10.3.2	Example 6: Means and Standard Deviations	95
10.4	Summary	96
10.5	Problems	97

11 Arrays 3: Further Examples — 98

11.1	Terminology	99
11.1.1	Rank	99
11.1.2	Bounds	99
11.1.3	Extent	99
11.1.4	Size	99
11.1.5	Shape	99
11.1.6	Conformable	99
11.2	Whole array manipulation	99
11.2.1	Assignment	100
11.2.2	Expressions	100
11.3	Array Sections	102
11.3.1	Example 1: Ages	102

11.3.2	Example 2: Examination Results	102
11.4	Allocatable Arrays	102
11.4.1	Example 3: Height Above Sea Level	103
11.5	Array Element Ordering	104
11.5.1	Array Element Ordering and Physical and Virtual Memory	105
11.6	Array Constructors	105
11.7	Masked Array Assignment and the WHERE Statement.	106
11.7.1	Notes	107
11.8	Summary	107
11.9	Problems	107

12 Output — 108

12.1	Integers, I format	109
12.2	Reals, F format	110
12.3	Reals, E format	111
12.4	Spaces	112
12.5	Alphanumeric or character format, A	113
12.6	Common mistakes	113
12.7	OPEN (and CLOSE)	114
12.7.1	The OPEN statement	114
12.7.2	Writing	115
12.8	Repetition	116
12.9	Some more examples	118
12.10	Implied DO loops	119
12.11	Formatting for a line-printer	120
12.11.1	Mechanics of carriage control	121
12.11.2	Generating a new line, on both line-printers and terminals.	122
12.12	Summary	123
12.13	Problems	123

13 Reading in Data — 125

13.1	Fixed fields on input	126
13.1.1	Integers, the I format	126
13.1.2	Reals, the F and E formats	126
13.2	Blanks, nulls and zeros	129
13.3	Characters	129
13.4	Skipping spaces and lines	130
13.5	Reading	130
13.6	File manipulation again	131
13.7	Errors when reading	132
13.8	Summary	132

13.9	Problems	132
14	**Functions**	**134**
14.1	An Introduction to Predefined Functions and Their Use	135
14.1.1	Example 1: Simple function usage	136
14.2	Generic Functions	136
14.2.1	Example 2: The ABS Generic function	136
14.3	Elemental Functions	137
14.3.1	Example 3: Elemental Function Use	137
14.4	Transformational Functions	137
14.4.1	Example 4: Simple Transformational Use	137
14.4.2	Example 5: Intrinsic DOT_PRODUCT use	138
14.5	Notes on Function Usage	138
14.6	Example 6: Easter	138
14.7	Complete List of Predefined Functions	140
14.7.1	Inquiry Functions	140
14.7.2	Transfer and Conversion Functions	141
14.7.3	Computational Functions	141
14.7.4	Array Functions	141
14.7.5	Pre-Defined Subroutines	142
14.8	Supplying your own functions	142
14.8.1	Example 7: Simple User Defined Function	142
14.9	An Introduction to the Scope of Variables and Local Variables	144
14.10	Recursive Functions	145
14.10.1	Example 8: Recursive Factorial Evaluation	145
14.11	Example 9: Recursive version of GCD	146
14.12	Example 10: After Removing Recursion	147
14.13	Internal functions	148
14.13.1	Example 11: Stirling's Approximation	148
14.14	Resume	149
14.15	Function Syntax	150
14.16	Rules and Restrictions	150
14.17	Problems	150
14.18	Bibliography	151
14.18.1	Recursion and Problem Solving	151
15	**Control Structures**	**152**
15.1	Selection between courses of action	153
15.1.1	The BLOCK IF statement.	153
15.1.2	Example 1: Quadratic Roots	156
15.1.3	Note	157
15.1.4	Example 2: Date calculation	157

15.1.5	The CASE Statement	158
15.1.6	Example 3: Simple calculator	158
15.1.7	Example 4: Counting Vowels, Consonants, etc	159
15.2	The three forms of the DO statement.	160
15.2.1	Example 5: Sentinel Usage	161
15.2.2	CYCLE and EXIT	162
15.2.3	Example 6: e**x Evaluation	162
15.2.4	Example 7: Wave Breaking on an Offshore Reef	163
15.3	Summary	165
15.4	Problems	166
15.5	Bibliography	169

16 Character — 170

16.1	Character Input	172
16.2	Character Operators	173
16.3	Character Sub-Strings	174
16.4	Character functions	176
16.5	Summary	178
16.6	Problems	178

17 Complex — 180

17.1	Example	182
17.2	Complex and Kind Type	183
17.3	Summary	183
17.4	Problems	183

18 Logical — 184

18.1	I/O	187
18.2	Summary	188
18.3	Problems	188

19 User Defined Types — 189

19.1	Example 1 – Dates	190
19.2	Type Definition	190
19.3	Variable Definition	191
19.4	Example 2 – Address lists	191
19.5	Example 3: Nested User Defined Types	192
19.6	Problems	194
19.7	Bibliography	194

20 Dynamic Data Structures — 195

20.1	Example 1: Simple Pointer Concepts	196
20.2	Example 2: Singly linked list	197

20.3	Other Dynamic Data Structures	199
20.4	Trees	200
20.4.1	Example 3: Perfectly Balanced Tree	200
20.5	Using Linked Lists for Sparse Matrix Problems	203
20.5.1	Inner Product of two Sparse Vectors	204
20.6	Data Structures Summary	208
20.7	Problems	208
20.8	Bibliography	209

21 Files — 211

21.1	Files in Fortran	212
21.2	Summary of options on OPEN	214
21.3	More fool proof i/o	215
21.4	Summary	217
21.5	Problems	217

22 An Introduction to Subroutines — 218

22.1	Simple Subroutine Example	219
22.2	Defining a subroutine	222
22.3	Referencing a subroutine	222
22.4	Dummy Arguments or Parameters, and Actual Arguments	222
22.5	Interface	222
22.6	Intent	223
22.7	Local Variables	224
22.7.1	Local Variables and the SAVE attribute	224
22.8	Scope of Variables	224
22.9	Status of the Action Carried out in the Subroutine	224
22.10	Why Bother?	225
22.11	Summary	225
22.12	Problems	225

23 Subroutines: 2 — 226

23.1	Example 1: Introduction to Arrays as Parameters	227
23.1.1	Explicit Shape Dummy Arrays	227
23.2	Example 2: Characters as parameters and assumed length dummy arguments	228
23.3	Example 3: Using Hoare's Quick Sort Algorithm	229
23.3.1	Note 2 – Intent Attribute	235
23.3.2	Note 3 – Explicit shape dummy array	235
23.3.3	Note 4 – Assumed Length Dummy Argument	235
23.3.4	Note 5 – Recursive Subroutine	235
23.3.5	Note 6 – Internal Subroutines and Scope	235
23.3.6	Note 7 – Flexible Design	236

23.4	Example 4: Rank two and higher arrays as parameters	236
23.4.1	Assumed Shape Arrays	236
23.5	Summary	238
23.6	Problems	238
23.7	Bibliography	239

24 An Introduction to Modules — 241

24.1	Modules for global data	242
24.1.1	Example 1: Modules for Precision Specification and Constant Definition	243
24.1.2	Note	243
24.1.3	Example 2: Constant Definition and Array Definition	244
24.2	Modules for derived data types	245
24.2.1	Example 3: Person Data Type	245
24.3	Modules for explicit procedures interfaces	248
24.3.1	Example: Using QuickSort	248
24.4	Modules containing procedures	249
24.5	Example 4 – The Solution of Linear Equations Using Gaussian Elimination	251
24.5.1	Notes	256
24.6	Notes on Module Usage and Compilation	256
24.7	Summary	257
24.8	Problems	257
24.9	Bibliography	258

25 Formal Syntax and Some Additional Features — 259

25.1	Program Units	260
25.2	Procedure – Function or Subroutine	260
25.2.1	Internal Procedure	260
25.3	Module	260
25.4	Executable Statements	261
25.5	Statement Ordering	261
25.6	Entities	262
25.7	Scope and Association	262
25.8	Modules and Scope	264
25.8.1	Public and Private Attributes	264
25.8.2	USE, ONLY and Rename	265
25.9	Keyword and Optional Arguments	266
25.10	Syntax Summary of Some Frequently used Fortran Constructs	267
25.10.1	Main Program	267
25.10.2	Subprogram	267
25.10.3	Module	268
25.10.4	Internal Procedure	268

25.10.5	Procedure heading	268
25.10.6	Procedure ending	268
25.10.7	Specification construct	268
25.10.8	Derived Type definition	268
25.10.9	Interface block	269
25.10.10	Specification statement	269
25.10.11	Type specification	269
25.10.12	Attribute Specification	269
25.10.13	Executable construct	270
25.10.14	Action statement	270

26 Case Studies — 271

26.1	Example 1 – Solving a System of First Order Ordinary Differential Equations using Runge-Kutta-Merson	272
26.1.1	Note: Alternative form of the Allocate statement	278
26.1.2	Note: Automatic arrays	278
26.1.3	Note: Dummy Procedure Arguments	279
26.2	Example 2 – Generic Procedures	279
26.3	Example 3 – A Function that returns a variable length array	285
26.4	Example 4 – Operator and Assignment Overloading	286
26.5	Example 5: A Subroutine to Extract the Diagonal Elements of a Matrix	287
26.6	Modules and Packaging	288
26.7	Problems	290
26.8	Bibliography	290

27 Converting from Fortran 77 — 291

27.1	Deleted Features	292
27.2	Obsolescent Features	292
27.2.1	Arithmetic IF	292
27.2.2	Real and Double precision DO Control Variables	292
27.2.3	Shared DO termination and non ENDDO termination	292
27.2.4	Alternate RETURN	292
27.2.5	PAUSE Statement	293
27.2.6	ASSIGN and assigned GOTO statements	293
27.2.7	Assigned FORMAT statements	293
27.2.8	H Editing	293
27.3	Better Alternatives	293

28 Miscellanea — 295

28.1	Program Development and Software Engineering	296
28.1.1	Modules	296
28.1.2	Programming Style – Programs should be Easy to Read	297
28.1.3	Programming Style – Programs should Behave Well	297

28.2	Data Structures	298
28.3	Algorithms	298
28.4	Recursion	298
28.5	Structured Programming and the GOTO Statement	298
28.6	Efficiency, space time trade off	299
28.7	Program Testing	299
28.8	Simple Debugging Techniques	300
28.9	Software Tools	300
28.9.1	Cross Referencing	300
28.9.2	Pretty print	300
28.9.3	NAGWare f90 Tools	300
28.10	Numerical Software Sources	300
28.11	Coda	301
28.12	Bibliography: All sources (bar one) taken from comp.software-eng.	302
28.12.1	Software Engineering	302
28.12.2	Programming Style	302
28.12.3	Software Testing	302
28.12.4	Fun	302
Glossary		303
Appendix A, Sample Program Examples		309
Appendix B, ASCII Character Set		312
Appendix C, Intrinsic Functions and Procedures		313
Appendix D, English and Latin Texts		340
Appendix E, Coded Text Extracts		341
Appendix F, NAG		342
Appendix G, Annex D, ISO/IEC 1539: 1991 (E)		344
Index		366

1
Overview

'I don't know what the language of the year 2000 will look like, but it will be called Fortran.'
C. A. R. Hoare

Aims

The aims of the chapter are to provide a background to the organisation of the book.

1 Overview

The book aims to provide coverage of a recommended subset of the full Fortran 90 language. The subset that is chosen is one that fits most closely with the theory and practice of structured programming, data structures and software engineering.

Chapters 2–5 provide a short background to computer systems and their use.

Chapter 2   This chapter looks at the basics of computer systems from the hardware point of view.

Chapter 3   This chapter provides a short history of operating system developments and looks at some commonly used operating systems.

Chapter 4   This chapter looks at some of the fundamentals of using a computer system.

These three chapters provide information that will be very useful in the longer term for the successful use of computer systems for programming.

Chapters five and six provide a coverage of problem solving and the history and development of programming languages. Chapter five is essential for the beginner as the concepts introduced here are used and expanded on throughout the rest of the book. Chapter six can be omitted initially but should be covered at some time. Programming languages evolve and some understanding of where Fortran has come from and where it is going will prove valuable in the longer term.

Chapter 5   This chapter looks at problem solving in some depth, and there is a coverage of the way we define problems, the role of algorithms, the use of both top-down and bottom up methods, and the requirement for formal systems analysis and design for more complex problems.

Chapter 6   This chapter looks at the history and development of programming languages. It helps to put many of the new features of Fortran 90 into context. The X3J3 Committee drew on a wide range of ideas for consideration for inclusion in Fortran 90. There is a brief coverage of the developments in the high performance Fortran area, which is likely to lead to changes to Fortran in 1995–1996. Example programs for a subset of the languages described are given in Appendix A.

Chapters seven through eleven cover the major features provided in Fortran for numeric programming in the first instance and for general purpose programming in the second. Each chapter has a range of problems. It is essential that a reasonable range of problems are attempted and completed. It is impossible to learn any language without practice.

Chapter 7   This chapter provides an introduction to programming with some simple Fortran 90 examples. For people with knowledge of programming this chapter can be covered fairly quickly.

Chapter 8   This chapter looks in more depth at arithmetic in some depth, with a coverage of the various numeric data types, expressions and assignment of scalar variables. There is also a thorough coverage of the facilities provided in Fortran 90 in helping to write programs that work on quite widely differing hardware.

Overview

Chapter 9 This chapter is an introduction to arrays and do loops. The chapter starts with some examples of tabular structures that one should be familiar with. There is then an examination of what concepts we need in a programming language to support manipulation of tabular data.

Chapter 10 Here we take the ideas introduced in the previous chapter and extend them to higher dimensioned arrays, additional forms of the DIMENSION attribute and corresponding form of the DO loop, and the use of looping for the control of repetition and manipulation of tabular information without the use of arrays.

Chapter 11 Here we look at more of the facilities offered for the manipulation of whole arrays and array sections, ways in which we can initialise arrays using constructors, look more formally at the concepts we need to be able to accurately describe and understand arrays, and finally look at the differences between the way Fortran allows to to use arrays and the mathematical rules governing matrices.

Chapters nine through eleven provide a coverage of some of the more important features and uses of arrays in the field of numerical problem solving. The framework provided here is drawn upon in later chapters in the book, with more complex and realistic examples.

Chapters twelve and thirteen look at input and output (I/O) in Fortran. An understanding of I/O is necessary for the development of so called production, non-interactive programs. These are essentially fully developed programs that are used repeatedly with a variety of data inputs and results.

Chapter 12 This chapter looks at output of results and how to generate something that is more comprehensible and easy to read than that available with free format output, and also how to write the results to a file rather than the screen.

Chapter 13 This chapter extends the ideas introduced in chapter 12 on output to cover input of data, or reading data into a program and considers file i/o also.

Chapter fourteen introduces the first building block available in Fortran for the construction of programs for the solution of larger more complex problems.

Chapter 14 This chapter looks at the functions available in Fortran, the so called intrinsic functions and procedures, over 100 of them, and also covers how you can define and use your own functions.

It is essential to develop an understanding of the functions provided by the language and when it is necessary to write your own.

Chapter fifteen introduces more formally the concept of control structures and their role in structured programming. Some of the control structures available in Fortran have been introduced in earlier chapters, but there is a summary here of those already covered plus several new ones that complete our coverage of a minimal working set.

Chapter 15 This chapter looks at the control structures available, from the viewpoint of established theory and practice in structured programming.

Chapters sixteen through twenty complete our coverage of the data typing and structuring facilities provided by Fortran, both pre-defined and user-defined. Fortran has now caught up with some of the major developments in the data structuring area of the last twenty years, and available in other languages for some time.

Chapter 16   This chapter looks at the character data type in Fortran. There is a coverage of I/O again, with the operators available – only one in fact.

Chapter 17   This chapter looks at the last numeric data type in Fortran, the complex data type. This data type is essential to the solution of a small class of problems in mathematics and engineering.

Chapter 18   Logical. The material covered here helps considerably in increasing the power and sophistication of the way we use and construct logical expressions in Fortran. This proves invaluable in the construction and use of logical expressions in control structures.

Chapters 19   User defined. This introduces another major new feature of Fortran. Previous versions of the language lacked any facilities here. This meant that in many applications earlier versions of Fortran were not the language of first choice for many people.

Chapter 20   This chapter looks at the dynamic data structuring facilities introduced in Fortran 90. Examples are drawn from a range of sources.

These chapters conclude coverage of the data structuring facilities provided by Fortran. There are problems that will require facilities not provided, but is surprising what can be achieved with the set now provided in Fortran. The material covered is extended into more realistic examples when we look the construction of larger and more complex programs in the last chapters in the book.

Chapter 21   Here we provide a more formal coverage of files, than provided in the earlier chapters on input and output.

Chapter 22–26   These chapters look at program structure, subroutines, procedures and modules, from the viewpoint of software engineering. Chapter 22 provides an gentle introduction to some of the fundamental concepts of subroutine definition and use; Chapter 23 extends these ideas. Chapter 24 introduces the concept of a module and the range of things that this brings to Fortran. Chapter 25 looks at a number of concepts more formally and thoroughly than earlier coverages. This chapter also provides a coverage of some of the more advanced features of Fortran 90. Chapter 26 has a small number of case studies helping to pull together the ideas presented in the earlier chapters.

Chapter 27   This chapter one looks at language features of Fortran 77 and how to convert (where necessary) to the newer and better features of Fortran 90.

Chapter 28   This provides an overview to the complete coverage of the material in the rest of the book. It looks at program development and software engineering, modules, programming style, data structures, algorithms, structured programming, recursion and recursion removal, efficiency in space and time, program testing, simple debugging techniques, software tools, numerical software sources.

Overview

Many of the chapters have an annotated bibliography. These provide pointers and directions for further reading. The coverage provided cannot be seen in isolation. The concepts introduced are by intention brief, and fuller coverages must be sought where necessary.

There is a glossary which provides coverage of both of the new concepts provided by Fortran 90 and a range of computing terms and ideas.

As mentioned earlier appendix A provides an example of a simple program in a number of the languages described in the chapter on program language development. There is also a coverage of the standards that apply.

Appendix B is the ASCII Character set.

Appendix C contains a list of all of the intrinsic procedures in Fortran 90. This contains a full explanation of each intrinsic procedure with a coverage of the rules and restrictions that apply and examples of use.

Appendix D contains the English and Latin text extracts used in one of the problems in the chapter on characters.

Appendix E contains the coded text extract used in one of the problems in the chapter on characters.

Appendix F provides a brief coverage of the NAG library.

Appendix G provides a coverage of Annex D of the Fortran 90 standard. This provides a formal definition of the whole language, with rules and restrictions. It cannot possibly be a substitute for the complete standard, but will provide a useful guide to the syntax and semantics of the language.

The book is not and cannot possibly be completely self contained and exhaustive in its coverage of the Fortran 90 language. Our first intention has been to produce a coverage of the features that we get you started with Fortran, and enable you to solve quite a wide range of problems successfully.

Fortran, like most languages, has features that are of relatively little use or make the construction of larger scale programs more difficult, especially when moving between hardware platforms. We have deliberately avoided these features.

Another problem is backwards compatibility with Fortran 77. Existing Fortran 77 programs have to be maintained, and there is much in that language that is *deprecated* or *obsolescent* in terms of Fortran 90.

We have aimed to introduce a working subset of the new language that emphasises the better constructs provided in Fortran 90 over its predecessors, Fortran 77 and Fortran 66.

All in all Fortran 90 is a quite exciting language, and has *caught up* with language developments of the 1970's and 1980's.

2
Introduction to Computer Systems

'Don't Panic'

Douglas Adams, *The Hitch-Hiker's Guide to the Galaxy*

Aims

The aims of this chapter are to introduce the following:-

- the components of a computer – the hardware;
- the component parts of a complete computer system – the other devices that you need to do useful work with a computer;
- the software needed to make the hardware do what you want;

2 Introduction to Computer Systems

A computer is an electronic device, and can be thought of as a tool, like the lever or the wheel, which can be made to do useful work. At the fundamental level it works with *bits* (binary digits or sequences of zeros and ones). Bits are generally put together in larger configurations, e.g. 8, 16, 32, or 64. Hence computers are often referred to as 8-bit, 16-bit, 32-bit, or 64-bit machines.

Most computers consist of the following:–

cpu
: This is the brains of the computer. CPU stands for *central processor unit*. All of the work that the computer does is organised here.

memory
: The computer will also have a memory. Memory on a computer is a solid state device that comprises an ordered collection of bits/bytes/words that can be read or written by the CPU. A byte is generally 8 bits (as in *8-bit byte*), and a word is most commonly accepted as the minimum number of bits that can be referenced by the CPU. This referencing is called *addressing*. The memory typically contains programs and data.

  A word size of 32 bits is the most common, and can be found in micro-computers, workstations, and many mini and mainframe computers. A word size of 64 bits is the domain of the most powerful workstation, main frame computers and super-computers.

  A computer memory is often called random access memory, or RAM. This simply means that the access time for any part of memory is the same; in order to examine location (say) 97, it is not necessary to first look through locations 1 to 96. It is possible to go directly to location 97. A slightly better term might have been *access at random*. The memory itself is highly ordered.

bus
: A *bus* is a set of connections between the CPU and other components. The bus will be used for a variety of purposes. These include address signals which tell the memory which words are wanted next; data lines which are used to transfer data to and from memory, and to and from other parts of the computer system. This is typical of many systems, but systems do vary considerably; while the information above may not be true in specific cases, it provides a general model.

2.1 The components of a computer system

So far the computer we have described is not sufficiently versatile. We have to add on other pieces of electronics to make it really useful.

Disks
: These are devices for storing collections of *bits*, which are inevitably organised in reality into bytes and files. One advantage of adding these to our computer system is that we can go away, switch the machine off, and come back at a later time and continue with what we were doing.

Memory is expensive and fast whereas disks are slower but cheaper. Most computer systems balance speed against cost, and have a small memory in relation to disk capacity.

Most people would be familiar with the two main type of disks on micro computers, and these are floppy disks, and hard disks. Micro floppy disks now come in one main physical size, 3 1/2 inch, though 5 1/4 inch was popular, and smaller sizes are also used. Hard disks are inside the system, and most people do not see this type of disk.

CD-ROMS are increasingly becoming popular on many systems.

Others   There are a large number of other input and output devices. These vary considerably from system to system depending on the work being carried out. These include printers of a variety of types, plotters, photo-type-setters, pens, sound interfaces, both for speech recognition and sound production, scanners, modem interfaces, fax cards, network interfaces etc.

The most important i/o devices are the keyboard and screen, whether you use a terminal, microcomputer or workstation. This book has been written assuming that most of your work will be done at one of these devices.

Terminals fall into two categories, character based devices (and the DEC VT series is a very popular one) and graphical devices (and the X-Windows terminals are the most popular). Terminal access to remote systems is often provided on micro-computers using terminal emulation software, e.g. Kermit, EMU-TEK, Vista eXceed.

Microcomputers provide the opportunity of cheap and powerful desktop computing facilities, where the processing is done locally.

Workstations are more powerful than microcomputers but this division is becoming rather blurred with the recent generations of processors. Screens on these devices are graphically orientated.

This means that the device we use looks rather like an ordinary typewriter keyboard, although some of the keys are different. However, the location of the letters, numbers and common symbols is fairly standard. Don't panic if you have never met a keyboard before. You don't have to know much more than where the keys are. Few programmers, even professionals, advance beyond the stage of using two index fingers and a thumb for their typing. You will find that speed of typing is rarely important, it's accuracy that counts.

One thing that people unfamiliar with keyboards often fail to realise is that what you have typed in is not sent to the computer until you press the *carriage return* key. To achieve any sort of communication you must press that key; it will be somewhere on the right hand side of the keyboard, and will be marked *return, c/r, send, enter*, or something similar.

2.2 Software

So far we have not mentioned software. Software is the name given to the programs that *run* on the hardware. Programs are written in *languages*. Computer languages are frequently divided into two categories; *high-level* and *low-level*. A low level language (e.g. assembler) is closer to the hardware, while a high level language (e.g. Fortran) is closer to the problem statement. There is typically a one to one correspondence between an assembly language statement and the actual hardware instruction. With a high level language there is a one to many correspondence; one high level statement will generate many machine level instructions.

A certain amount of general purpose software will have been provided by the manufacturer. This software will typically include the basic operating system, one or more *compilers*, an *assembler*, an *editor*, and a *loader* or *link editor*.

- A *compiler* translates high level statements into machine instructions;
- An *assembler* translates low level or assembly language statements into machine instructions;
- An *editor* makes changes to text files, e.g. program sources;
- A *loader* or *link editor* takes the output from the compiler and completes the process of generating something that can be executed on the hardware.

These programs will vary considerably in size and complexity. Certain programs that make up the operating system will be quite simple and small (like copying utilities), while certain others will be relatively large and complex (like a compiler).

In this book we concentrate on software or programs that you write for your research, work or course work. As the book progresses you will be introduced to ways of building on what other people have written, and how to take advantage of the vast amount of software already written, tested and documented.

2.3 Problems

1. Distinguish between a memory address and memory contents.

2. What does RAM stand for?

3. What would a WOM (write only memory) do? How would you use it?

4. What does CPU stand for? What does it do?

5 What does a compiler do?

6 What does a linker do?

2.4 Bibliography

Baer J.L., *Computer Systems Architecture*, Computer Science Press.

- Extremely readable coverage of this whole area. The version could do with an update, but it is still a very impressive coverage. Highly recommended

Hayes J.P., *Computer Architecture and Organisation*, McGraw Hill.

- Provides coverage of the evolution of computers, including a short section on the mechanical era, an area not covered in all books. There is also coverage of design methodology, processor design, control design, memory organisation and system organisation.

Hamacher V.C., Vranesic Z.G., Zaky S.G., *Computer Organisation*, McGraw Hill.

- Provides coverage of address methods, machine program sequencing, instructions sets and their implementation, the CPU, microprogramming, i/o organisation, arithmetic, memory, peripherals and interfacing, software, microprocessors, communications. Quite easy to read.

Intel, *i486 Microprocessor Hardware Reference Manual,* Intel.

- Most if not all processor manufacturers provide publications that describe their hardware. The Intel family of processors based round the original 8086 now offers quite high performance, and is commonly found in desk top machines. Appendix A provides a good introduction to the Intel 86 family architecture, both for people who have used systems based on the original IBM PC since the outset, and those using systems for the first time today, using 80486 and Pentium based systems.

Reeves C.M., *An Introduction to Logical Design of Digital Circuits*, CUP.

- This book provides coverage of the construction of the very simple electronic building blocks, from which most modern computer systems are made of. Relatively theoretical.

Tannenbaum A.S., *Structured Computer Organisation*, Prentice Hall.

- Very good coverage looking at a computer system in terms of a hierarchy of levels. An easy read.

3
Introduction to Operating Systems

'Where shall I begin your Majesty' he asked.
'Begin at the beginning,' the King said, gravely 'and go on till you come to the end then stop.'
Lewis Carroll, *Alice's Adventures in Wonderland.*

Aims

The aims of this chapter are to introduce the following

- to provide a brief history of operating system development
- to look briefly at some commonly used operating systems;
    - DOS and Windows
    - UNIX and X Windows
    - VMS and Open VMS

3 Introduction to Operating Systems

A simple definition of an operating system is the suite of programs that make the hardware usable. Most computer systems provide one. They vary considerably from those available on early microcomputers, like CP/M, to DOS on PCs, UNIX on workstations, to very complex and powerful ones on mainframes, e.g. IBMs VM/CMS. UNIX is also a popular choice of operating system for super computers.

From the designers point of view they are mainly resource manages. They allow management of the CPU, disks and i/o devices. They have to provide a user interface for computer operators, professional programmers (whether systems or applications), administrators of the system, and finally the casual end user. As can be imagined, these groups have different functional requirements. It is therefore useful to look at the development of operating systems, and see a shift from satisfying the requirements of the professional to satisfying the requirements of the casual end user.

3.1 History of Operating Systems

3.1.1 The 1940s

Early computer systems had no operating systems in the modern sense of the word. An early commercially available machine was the IBM 604. This machine could undertake some 60 program steps before using punch cards as backing store. The end user had intimate knowledge of the machine and programmed at a very low level.

3.1.2 The 1950s

This era saw a rapid change in the capabilities of operating systems. They were designed to make efficient use of an expensive resource. Jobs were *batched* so that the time between jobs was minimised. The end user was now distanced from the machine. This era saw rapid development in program language design and a notable end to the period was the design of Algol 60.

3.1.3 The 1960s

The next milestone was the introduction of multiprogramming. Probably initially seen as a way of making efficient use of hardware it heralded the idea of time sharing. A time sharing systems is characterised by the conversational nature of the interaction, and the use of a keyboard. This had a tremendous impact on the range of uses that a computer system had and on the program development process.

3.1.4 The 1960s and 1970s

The realisation that computer systems could be used in a large range of human activities saw the development of large general purpose systems, and probably the most famous of these was the IBM 360 series. These systems were some of the most complex programming endeavours undertaken, and most projects were late and well over budget. These costly mistakes helped lead to the establishment of software engineering as a discipline.

The contribution of the time sharing system to the program development was quickly realised to be considerable. A system that was developed during this period was UNIX – and this operating system has a very sharp set of tools to aid in program development.

3.1.5 The 1970s and 1980s

The 70s represented a period of relative stability with newer and more complex versions of existing systems.

During this period the importance of graphical interfaces emerged and with dropping hardware costs graphical interfaces started to dominate.

The Apple Macintosh heralded a new era and became a popular choice for many people. At the same time graphical interfaces were being added to existing major operating systems, with X-Windows and associated higher level systems hiding raw UNIX from many users, and Windows being added as an option to DOS on the Intel family of processors.

UNIX System 5 Release 4 is likely to become the standard in the near future, with appropriate standard graphical interfaces sitting on top, and enabling the end user to buy whatever hardware they want and still have a consistent interface.

Windows NT is likely to become a major industry standard operating systems, under development by the Microsoft organisation, and being chosen by a number of other companies also, e.g. DEC on the new Alpha processor.

3.2 Networking

Networking simplistically is a way of connecting two or more computer systems, and networking computer systems is not new, and one of the first was the SAGE military network, funded by the US DoD in the 1950's.

Networking capability has undergone a massive increase during the computer age. Local networks of two or three systems, through tens of systems in small research groups and organisation are now extremely common place. It is not uncommon now to have in excess of a thousand network connected devices on one local area network.

Wide area networking is also quite common, and most major organisations now have networks spanning a country or even the whole world.

One of the most widely used wide area networks in the academic and scientific world is the INTERNET, and there are over thirty million systems on the INTERNET at the time of writing this book.

A number of books on networking are included in the bibliography.

3.3 Problems

1. What type of system do you use, i.e. is it a stand alone micro computer, terminal, work station etc?

2. Is it networked, and if so in what way?

3. Is wide area networking available?

4. Is a graphical interface available?

3.4 Bibliography

Blomeley F., *Networks and Network Services: A User's Guide*, Im**media**te Publishing.

- Very readable perspective of the Internet, from a British view point, and hence of considerable interest to the academic community in Britain currently moving from an X25 based system to an IP based system.

Brooks F. P., *The Mythical Man Month: Essays on Software Engineering,* Addison Wesley.

- A very telling coverage of the development of the operating system for the IBM 360 series of systems. A must for any one involved in the longer term in program development.

Deitel H.M., *An Introduction to Operating Systems*, Addison Wesley.

- One of the most accessible books on operating systems with coverage of process management, storage management, processor management, auxiliary storage management, performance, networks and security, with case studies of the major players including UNIX, VMS, CP/M, MVS, VM, DOS and Windows.

Feit S., *TCP/IP, Architecture, Protocols, and Implementation,* McGraw Hill.

- A more technical book than Blomeley or Kroll, well written with a wealth of information for the more inquisitive reader.

Kroll E., *The Whole Internet User's Guide and Catalog,* O'Reilly.

- **The** Internet book. Written with very obvious enthusiasm by Mister Internet himself!

4
Introduction to Using a Computer System

'Maybe one day we shall be glad to remember even these things.'
Virgil.

Aims

The aims of this chapter are to introduce some of the fundamentals of using a computing system, including

- files
- editors
- systems access and networking

4 Introduction to Using a Computer System

There are a number of concepts that underpin your use of any computing system. Sitting at a high resolution colour screen with a miriad of icons this may not be immediately apparent, but developing an appreciation of this will help considerably in the long term, and when inevitably moving from one system to another.

4.1 Files

A file is a collection of information that you refer to by name, e.g. if you were to use a word processor to prepare a letter then the letter would exists independently as a file on that system, generally on a disk. With graphical interfaces there will be a systematic iconic representation of files.

There will be many ways of manipulating files on the operating system that you work on. You will use an editor to make changes to a file. This file might be the source of a program and you could then use a compiler to compile your program. The compiler will generate a number of files some of interest to you, others for its own use. There will be commands in the operating system to make copies of files, back files up onto a variety of media etc.

4.2 Editors

All general purpose computer systems have at least one editor so that you can modify programs and data. Screen editors are by far the most easy to use, with changes you make to the file being immediately visible on the screen in front of you.

Some editors will have sophisticated command modes of operation with pattern matching allowing very powerful text processing capability. These can automate many common tasks, taking away the manual, repetitive drudgery of screen based editing.

4.3 Stand Alone Systems

These are becoming increasingly common, in use both at work and in the home. The IBM pc and compatibles is a very popular choice in the scientific community. They offer ease of use and access to a considerable amount of raw processing power for compute intensive applications.

4.4 Networked Systems

It is quite common to interconnect the above in to local area networks. This same network would also have file servers, printers, plotters, mail gateways etc.

Work stations are generally networked in an environment like the above, providing very powerful processing capability.

4.5 Multi-User Systems

One step above micro-computers and work stations are multi-user systems. The dividing line between micro computers, individual work stations and multi-user systems is rapidly becoming blurred.

Multi-user systems, especially the larger ones, are very popular as they take away from the casual end user much of the drudgery of the day to day tasks of backing up disks, installing new versions of the software, locating and fixing problems with software that doesn't work quite as it should etc to a system manager or operator.

Here we find one person with the role of registering new users, backing up the file system, sorting out printer problems, networking problems etc. This also means that all of the users of the system do not have to remember rather arcane and sometimes rather magical commands! They can get on with solving their actual problem.

4.6 Other Useful Things to Know

You will soon need to know what files you are working with and there will be commands to do this. There will be a need to get rid of files and there will be commands to achieve this.

There will be ways of getting on-line help, and *help* as a command is (for once!) used by a variety of operating systems. On UNIX systems the rather more unintelligible *man* command is available.

There will be commands to print program listings and data files

With networked and multi user systems there will be commands to send and receive electronic mail to/from other users. It is easy to send and reply to mail from people across the world, often in hours and even minutes. The table below has examples of some common operating system commands in DOS, UNIX and VMS.

Operating System and Command	DOS	UNIX	VMS
What files are there	dir	ls	dir
Get rid of a file	del	rm	del
Copy a file	copy	cp	copy
Display a file on the screen	type	cat	type
Print a file	print	pr	print
Create or make changes to a file	edit	ed	edit
		vi	edit/tpu
Make a sub-directory	mkdir, md	md	create/dir
Change to another directory	chdir, cd	cd	set default

Chapter 6, section 6.16.3, provides a variety of sources of information regarding Fortran and Fortran 90. These include things like language sensitive editors and tools for programming.

4.7 Bibliography

The main source here are the manuals and documentation provided by the supplier of whatever system you use. These are increasingly of a very high standard. However inevitably written to highlight the positive and down play the negative aspects of the systems.

The next source are the many third party books written and widely available throughout the world. These vary in price considerably from basic introductory coverages to very comprehensive reference works. These are a very good complement to the first.

Gilly D., *UNIX in a Nutshell,* O'Reilly and Associates.

- A very good quick reference guide. Assumes some familiarity with UNIX. Current edition (at the time of writing this book) was System V Release IV, with Solaris 2.0. Also provides coverage of the various shells, Bourne, Korn and C.

Livingston B., *Windows 3.1 Secrets*, IDG Books.

- Very good coverage of Windows 3.1. Suitable for both the beginner and more knowledgeable user. Whilst Microsoft have improved their documentation considerably, and the current documentation is of a high standard, this is a must for serious Windows users.

Microsoft, *MS-DOS 6 User Guide*, Microsoft Press.

- Good introduction to DOS. Suitable for beginners to intermediate level user.

Microsoft, *Windows User's Guide*, Microsoft Press.

- Good coverage of Windows and suitable for the beginner and intermediate level user. Sufficient for most users. A massive improvement over earlier versions.

5
Introduction to Problem Solving

They constructed ladders to reach to the top of the enemy's wall, and they did this by calculating the height of the wall from the number of layers of bricks at a point which was facing in their direction and had not been plastered. The layers were counted by a lot of people at the same time, and though some were likely to get the figure wrong the majority would get it right... Thus, guessing what the thickness of a single brick was, they calculated how long their ladder would have to be.

Thucydides, *The Peloponnesian War*

'When I use a word,' Humpty Dumpty said, in a rather scornful tone, 'it means just what I choose it to mean - neither more nor less'
'The question is,' said Alice, 'whether you can make words mean so many different things.'

Lewis Carroll, *Through the Looking Glass and What Alice found there.*

Aims

The aims are:–

- to examine some of the ideas and concepts involved in problem solving;
- to introduce the concept of an algorithm;
- to introduce two ways of approaching algorithmic problem solving;
- to introduce the ideas involved with systems analysis and design, i.e. to show the need for pencil and paper study before using a computer system.

5 Introduction to Problem Solving

It is informative to consider some of the dictionary definitions of problem

- a matter difficult of settlement or solution, *Chambers*
- a question or puzzle propounded for solution, *Chambers*
- a source of perplexity, *Chambers*
- doubtful or difficult question, *Oxford*
- proposition in which something has to be done, *Oxford*
- a question raised for enquiry, consideration, or solution, *Webster's*
- an intricate unsettled question, *Webster's*

and a common thread seems to be a question that we would like answered or solved. So one of the first things to consider in problem solving is how to pose the problem. This is often not as easy as is seems. Two of the most common methods are

- in natural language
- in artificial language or stylised language

Both methods have their advantages and disadvantages.

5.1 Natural Language

Most people use natural language and are familiar with it, and the two most common forms are the written and spoken word. Consider the following language usage

- the difference between a three year old child and an adult describing the world;
- the difference between the way an engineer and a physicist would approach the design of a car engine;
- the difference between a manager and worker considering the implications of the introduction of new technology;

Great care must be taken when using natural language to define a problem and a solution. It is possible that people use the same language to mean completely different things, and one must be aware of this when using natural language whilst problem solving.

Natural language can also be ambiguous

Old men and women eat cheese.

Are both the men and women old?

Introduction to Problem Solving

5.2 Artificial Language

The two most common forms of artificial language are technical terminology and notations. Technical terminology generally includes both the use of new words and alternate use of existing words. Consider some of the concepts that are useful when examining the expansion of gases in both a theoretical and practical fashion

- temperature
- pressure
- mass
- isothermal expansion
- adiabatic expansion

Now look at the following

- a chef using a pressure cooker
- a garage mechanic working on a car engine
- a doctor monitoring blood pressure
- an engineer designing a gas turbine

Each has a particular problem to solve, and each will approach their problem in their own way; thus they will each use the same terminology in slightly different ways.

5.2.1 Notations

Some examples of notations are

- algebra
- calculus
- logic

All of the above have been used as notations for describing both problems and their solutions.

5.3 Resume

We therefore have two ways of describing problems and they both have a learning phase until we achieve sufficient understanding to use them effectively. Having arrived at a satisfactory problem statement we next have to consider how we get the solution. It is here that the power of the algorithmic approach becomes useful.

5.4 Algorithms

An algorithm is a sequence of steps that will solve part or all of a problem. One of the most easily understood examples of an algorithm is a recipe. Most people have done some cooking, if only making toast and boiling an egg.

A recipe is made up of two parts

- a check list of things you need
- the sequence or order of steps

Problems can occur at both stages, e.g. finding out halfway through the recipe that you do not have an ingredient or utensil; finding out that one stage will take an hour when the rest will be ready in ten minutes. Note that certain things can be done in any order – it may not make any difference if you prepare the potatoes before the carrots.

There are two ways of approaching problem solving when using a computer. They both involve *algorithms*, but are very different from one another. They are called *top-down* and *bottom-up*.

5.4.1 Top Down

In a *top down* approach the problem is first specified at a high or general level: prepare a meal. It is then refined until each step in the solution is explicit and in the correct sequence, e.g. peel and slice the onions, then brown in a frying pan before adding the beef. One drawback to this approach is that it is very difficult to teach to beginners because they rarely have any idea of what *primitive* tools they have at their disposal. Another drawback is that they often get the sequencing wrong, e.g. *now place in a moderately hot oven* is frustrating because you may have not lit the oven (sequencing problem) and secondly because you may have no idea how hot *moderately hot* really is. However as more and more problems are tackled top-down becomes one of the most effective methods for programming.

5.4.2 Bottom up

Bottom-up is the reverse to top-down! As before you start by defining the problem at a high level, e.g. prepare a meal. However, now there is an examination of what tools etc you have available to solve the problem. This method lends itself to teaching since a repertoire of tools can be built up and more complicated problems can be tackled. Thinking back to the recipe there is not much point trying to cook a six course meal if the only thing that you can do is boil an egg and open a tin of beans. The bottom-up approach thus has advantages for the beginner. However there may be a problem when no suitable tool is present. One of the authors' friend's learned how to make Bechamel sauce, and was so pleased by his success that every other meal had a course with a bechamel sauce. Try it on your eggs one morning. Here was a case of specifying a problem *prepare a meal*, and using an inappropriate but plausible tool *Bechamel Sauce*.

The effort involved in tackling a realistic problem, introducing the constructs as and when they are needed and solving it is considerable. This approach may not lead to a reasonably comprehensive coverage of the language, or be particularly useful from a teaching point of view. Case studies do have great value, but it helps if you know the elementary rules before you start on them. Imagine learning French by studying Balzac, before you even look at a French grammar. You can learn this way but even when you have finished, you may not be able to speak to a Frenchman and be understood. A good example of the case study approach is given in the book *Software Tools*, by Kernighan and Plauger.

In this book our aim is to gradually introduce more and more tools until you know enough to approach the problem using the top-down method, and also realise from time to time that it will be necessary to develop some new tools.

5.4.3 Stepwise Refinement

Both the above techniques can be combined with what is called *step-wise refinement*. The original ideas behind this technique are well expressed in a paper by Wirth entitled *Program Development by Stepwise Refinement*, published in 1971. This means that you start with a global problem statement and break the problem down in stages, into smaller and smaller sub-problems, that become more and more amenable to solution. When you first start programming the problems you can solve are quite simple, but as your experience grows you will find that you can handle more complex problems.

When you think of the way that you solve problems you will probably realise that, unless the problem is so simple that you can answer it straight away some thinking and pencil and paper work is required. An example that some may be familiar with is in practical work in a scientific discipline, where coming unprepared to the situation can be very frustrating and unrewarding. It is therefore appropriate to look at ways of doing analysis and design before using a computer.

5.5 Systems Analysis and Design

When one starts programming it is generally not apparent that one needs a methodology to follow to become successful as a programmer. This is generally because the problems are reasonably simple, and it is not necessary to make explicit all of the stages one has gone through in arriving at a solution. As the problems become more complex it is necessary to become more rigorous and thorough in ones approach, to keep control in the face of the increasing complexity and to avoid making mistakes. It is then that the benefit of systems analysis and design becomes obvious. Broadly we have the following stages in systems analysis and design

- Problem definition
- Feasibility study and fact finding
- Analysis
- Initial system design
- Detailed design
- Implementation

- Evaluation
- Maintenance

and each problem we address will entail slightly different time spent in each of these stages. Let us look at each stage in more detail.

5.5.1 Problem Definition

Here we are interested in defining what the problem really is. We should aim at providing some restriction on both the scope of the problem, and the objectives we set ourselves. We can use the methods mentioned earlier to help out. It is essential that the objectives are

- clearly defined;
- when more than one person is involved, understood by all people concerned, and agreed by all people concerned;
- realistic.

5.5.2 Feasibility Study and Fact Finding

Here we look to see if there is a feasible solution. We would try and estimate the cost of solving the problem and see if the investment was warranted by the benefits, i.e. cost benefit analysis.

5.5.3 Analysis

Here we look at what must be done to solve the problem. Note we are interested in finding what we need to do, but we do not actually do it at this stage.

5.5.4 Design

Once the analysis is complete we know what must be done, and we can proceed to the design. We may find there are several alternatives, and we thus examine alternate ways in which the problem can be solved. It is here that we use the techniques of top-down and bottom-up problem solving, combined with step-wise refinement to generate an algorithm to solve the problem. We are now moving from the logical to the physical side of the solution. This stage ends with a choice between one of several alternatives. Note that there is generally not one ideal solution, but several, each with their own advantages and disadvantages.

5.5.5 Detailed Design

Here we move from the general to the specific, The end result of this stage should be a sufficiently tightly defined specification to generate actual program code from.

It is at this stage that it is useful to generate *pseudo-code*. This means writing out in detail the actions we want carried out at each stage of our overall algorithm. We gradually expand each stage (step-wise refinement) until it becomes Fortran – or whatever language we want in fact.

5.5.6 Implementation

It is at this stage that we actually use a computer system to create the program(s) that will solve the problem. It is here that we actually need to know sufficient about a programming language to use it effectively to solve our problems. This is only one stage in the overall process, and mistakes at any of the stages can create severe difficulties.

5.5.7 Evaluation and testing

Here we try to see if the program(s) we have produced actually do what they are supposed to. We need to have data sets that enable us to say with confidence that the program really does work. This may not be an easy task, as quite often we only have numeric methods to solve the problem, which is why we are using the computer to solve the problem — hence we are relying on the computer to provide the proof; i.e. we have to use a computer to determine the veracity of the programs – and as Heller says *Catch 22*.

5.5.8 Maintenance

It is rare that a program is run once and thrown away. This means that there will be an on going task of maintaining the program, generally to make it work with different versions of the operating system, compiler, and to incorporate new features not included in the original design. It often seems odd when one starts programming that a program will need maintenance as we are reluctant to regard a program in the same way as a mechanical object like a car that will eventually fall apart through use. Thus maintenance means keeping the program working at some tolerable level, with often a high level of investment in manpower and resources. Research in this area has shown that anything up to 80% of the manpower investment in a program can be in maintenance.

5.6 Conclusions

A drawback, inherent in all approaches to programming, and to problem solving in general, is the assumption that a solution is indeed possible. There are problems which are simply insoluble – not only problems like balancing a national budget, weather forecasting for a year, or predicting which radioactive atom will decay, but also problems which are apparently computationally solvable.

Knuth gives the example of a chess problem – determining whether the game is a forced victory for white. Although there is an algorithm to achieve this, it requires an inordinately large amount of time to complete. For practical purposes it is unsolvable.

Other problems can be shown mathematically to be undecidable. The work of Godel in this area has been of enormous importance, and the bibliography contains a number of references for the more inquisitive and mathematically orientated reader. The Hofstader coverage is the easiest, and least mathematical.

As far as possible we will restrict ourselves to solvable problems, like learning a programming language.

Within the formal world of Computer Science our description of an algorithm would be considered a little lax. For our introductory needs it is sufficient, but a more rigorous approach is given by Hopcroft and Ullman in *Introduction to Automata Theory, Languages and Computation*, and by Beckman in *Mathematical Foundations of Programming*.

5.7 Problems

1. What is an algorithm?

2. What distinguishes top-down from bottom-up approaches to problem solving? Illustrate your answer with reference to the problem of a car, motor-cycle or bicycle having a flat tire.

5.8 Bibliography

Aho A. V., Hopcroft J. E., Ullman J. D., *The Design and Analysis of Computer Algorithms*, Addison Wesley, 1982.

- Theoretical coverage of the design and analysis of computer algorithms.

Beckman F. S., *Mathematical Foundations of Programming*, Addison Wesley, 1981

- Good clear coverage of the theoretical basis of computing.

Bulloff J.J., Holyoke T.C., Hahn S.W., *Foundations of Mathematics – Symposium Papers Commemorating the 60th Birthday of Kurt Godel*, Springer Verlag.

- The comment by John von Neumann highlights the importance of Godel's work, .. *Kurt Godel's achievement in modern logic is singular and monumental – indeed it is more than a monument, it is a landmark which will remain visible far in space and time. Whether anything comparable to it has occurred in the logic of modern times may be debated. In any case, the conceivable proxima are very, very few. The subject of logic has certainly changed its nature and possibilities with Godel's achievement.*

Dahl O. J., Dijkstra E. W., Hoare C. A. R., *Structured Programming*, Academic Press, 1972.

- This is the seminal book on structured programming.

Davis M., *Computability and Unsolvability*, Dover, 1982.

- The book is an introduction to the theory of computability and non-computability – the theory of recursive functions in mathematics. Not for the mathematically faint hearted!

Davis W. S., *Systems Analysis and Design*, Addison Wesley, 1983.

- Good introduction to systems analysis and design, with a variety of case studies. Also looks at some of the tools available to the systems analyst.

Fogelin R. J., *Wittgenstein*, Routledge and Kegan Paul, 1980.

- The book provides a gentle introduction to the work of the philosopher Wittgenstein, who examined some of the philosophical problems associated with logic and reason.

Godel K., *On Formally Undecidable Propositions of Principia Mathematica and Related Systems*, Oliver and Boyd.

- An English translation of Godel's original paper by Meltzer, with quite lengthy introduction by R. B. Braithwaite, then Knightbridge Professor of Moral Philosophy at Cambridge University, England, and classified under philosophy at the library at King's, rather than mathematics.

Hofstader D.,*The Eternal Golden Braid*, Harvester Press.

- A very readable coverage of paradox and contradiction in art, music and logic, looking at the work of Escher, Bach and Godel respectively.

Hopcroft J. E., Ullman J. D., *Introduction to Automata Theory, Languages and Computation*, Addison Wesley, 1979.

- Comprehensive coverage of the theoretical basis of computing.

Kernighan B. W., Plauger P. J., *Software Tools,* Addison Wesley, 1976.

- Interesting essays on the program development process, originally using a non-standard variant of Fortran. Also available using Pascal.

Knuth D. E., *The Art of Computer Programming*, Addison Wesley,

Vol 1. *Fundamental Algorithms*, 1974

Vol 2. *Semi-numerical Algorithms*, 1978

Vol 3. *Sorting and Searching*, 1972

- Contains interesting insights into many aspects of algorithm design. Good source of specialist algorithms, and Knuth writes with obvious and infectious enthusiasm (and erudition).

Millington D., *Systems Analysis and Design for Computer Applications*, Ellis Horwood, 1981.

- Short and readable introduction to systems analysis and design.

Wirth N., *Program Development by Stepwise Refinement*, Communications of the ACM, April 1971, Volume 14, Number 4, pp. 221-227.

- Clear and simple exposition of the ideas of stepwise refinement.

6
Introduction to Programming Languages

'We have to go to another language in order to think clearly about the problem.'
Samuel R. Delany, *Babel–17*

Aims

The primary aim of this chapter is to provide a short history of program language development and provide some idea as to the concepts that have had an impact on Fortran 90. It concentrates on some but not all of the major milestones of the last 40 years, in rough chronological order. The secondary aim is to show the breadth of languages available. The chapter concludes with coverage of a small number of more specialised languages.

6 Introduction to Programming Languages

It is important to realise that programming languages are a recent invention. They have been developed over a relatively short period – 40 years, and are still undergoing improvement. Time spent gaining some historical perspective will help you understand and evaluate future changes. This chapter starts right at the beginning and takes you through some, but not all, of the developments during this 40 year span. The bulk of the chapter restricts itself to languages that are reasonably widely available commercially, and therefore ones that you are likely to meet. The chapter concludes with a coverage of some more specialised and/or recent developments.

6.1 Some Early Theoretical Work

Some of the most important early theoretical work in computing was that of *Turing* and *von Neumann*. Turing's work provided the base from which it could be shown that it was possible to get a machine to solve problems. The work of von Neumann added the concept of storage and combined with Turing's work to provide the basis of most computers designed to this day.

6.2 What is a programming language ?

For a large number of people a programming language provides the means of getting a digital computer to solve a problem. There are a wide range of problems, and an equally wide range of programming languages, with particular programming languages being suited to a particular class of problems. Thus there are a wide variety of programming languages, which often appears bewildering to the beginner.

6.3 Program Language Development and Engineering

There is much in common between the development of programming languages and the development of anything from the engineering world. Consider the car: old cars offer much of the same functionality as modern ones, but most people prefer driving newer ones. The same is true of programming languages, where you can achieve much with the older languages, but the newer ones are easier to use.

6.4 The Early Days

A concept that proves very useful when discussing programming languages is that of the level of a machine. By this is meant how close a language is to the underlying machine that the program *runs* on. In the early days of programming (up to 1954) there were only two broad categories, machine languages and assemblers. The language that a digital machine uses is that of 0 and 1, i.e. they are binary devices. Writing a program in terms of patterns of 0 and 1 was not particularly satisfactory and the capability of using more meaningful mnemonics was soon introduced. Thus it was realised quite quickly that one of the most important aspects of programming languages is that they have to be read and understood by *both* machines and humans.

6.4.1 Fortran

The next stage was the development of higher level languages. The first of these was *Fortran* and it was developed over a three year period from 1954 to 1957 by an IBM team lead by John Backus. This group achieved considerable success, and helped to prove that the way forward lay with high level languages for computer based problem solving. Fortran stands for *for*mula *tran*slation and was used mainly by people with a scientific background for solving problems that had a significant arithmetic content. It was thus relatively easy, for the time, to express this kind of problem in Fortran.

By 1966 and the first standard Fortran was

- widely available
- easy to teach
- had demonstrated the benefits of subroutines and independent compilation
- was relatively machine independent
- and often had very efficient implementations

Possibly the single most important fact about Fortran was and still is its widespread usage in the scientific community.

6.4.2 Cobol

The business world also realised that computers were useful and several languages were developed including FLOWMATIC, AIMACO, Commercial Translator and FACT, leading eventually to Cobol - *Co*mmon *B*usiness *O*rientated *L*anguage. There is a need in commercial programming to describe data in a much more complex fashion than for scientific programming, and thus Cobol had far greater capability in this area than Fortran. The language was unique at the time in that a group of competitors worked together with the objective of developing a language that would be useful on machines used by other manufacturers.

The contributions made by Cobol include

- firstly the separation between
    - the task to be undertaken
    - the description of the data involved
    - the working environment in which the task is carried out
- secondly a data description mechanism that was largely machine independent
- thirdly its effectiveness for handling large files
- fourthly the benefit to be gained from a programming language that was easy to read

Modern developments in computing, of report generators, file handling software, fourth generation development tools, and especially the increasing availability of commercial relational database management systems are gradually replacing the use of Cobol, except where high efficiency and or tight control are required.

6.4.3 Algol

Another important development of the 1950's was Algol. It had a history of development from Algol 58, the original Algol language, through Algol 60 eventually to the Revised Algol 60 Report. Some of the design criteria for Algol 58 were

- the language should be as close as possible to standard mathematical notation and should be *readable* with little further explanation
- it should be possible to use it for the description of computing processes in publications
- the new language should be mechanically translatable into machine programs

A sad feature of Algol 58 was the lack of any input/output facilities, and this meant that different implementations often had incompatible features in this area.

The next important step for Algol occurred at a UNESCO sponsored conference in June 1959. There was an open discussion on Algol and the outcome of this was Algol 60, and eventually the Revised Algol 60 Report.

It was at this conference that John Backus gave his now famous paper on a method for defining the syntax of a language, called Backus Normal Form, or BNF. The full significance of the paper by Backus was not immediately recognised. However BNF was to prove of enormous value in language definition, and helped provide an interface point with computational linguistics.

The contributions of Algol to program language development include

- block structure
- scope rules for variables because of block structure
- the BNF definition by Backus – most languages now have a formal definition
- the support of recursion
- its offspring

and thus Algol was to prove to make a contribution to programming languages that was never reflected in the use of Algol 60 itself, in that it has been the parent of one of the main strands of program language development.

6.5 Chomsky and Program Language Development

Programming languages are of considerable linguistic interest, and the work of Chomsky in 1956 in this area was to prove of inestimable value. Chomsky's system of transformational grammar was developed in order to give a precise mathematical description to certain aspects of language. Simplistically Chomsky describes grammars and these grammars in turn can be used to define or generate corresponding kinds of languages. It can be shown that for each type of grammar and language there is a corresponding type of machine. It was quickly realised that there was a link with the earlier work of Turing.

This link helped provide a firm scientific base for programming language development, and modern compiler writing has come a long way from the early work of Backus and his team at IBM. It may seem important when playing a video game at home or in an arcade, but for some it is very comforting that there is a firm theoretical basis behind all that fun.

6.6   Lisp

There were developments in very specialised areas also. List processing was proving to be of great interest in the 50's and saw the development of IPLV between 1954 and 1958. This in turn lead to the development of Lisp at the end of the 50's. It has proved to be of considerable use for programming in the areas of artificial intelligence, playing chess, automatic theorem proving and general problems solving. It was one of the first languages to be interpreted rather than compiled. Whilst interpreted languages are invariably slower and less efficient in their use of the underlying computer system, than compiled languages, they do provide great opportunities for the user to explore and try out ideas whilst sat at a terminal. The power that this gives to the computational problem solver is considerable.

Possibly the greatest contribution to program language development made by Lisp was its functional notation. One of the major problems for the Lisp user has been the large number of Lisp flavours, and this has reduced the impact that the language has had and deserved.

6.7   Snobol

Snobol was developed to aid in string processing which was seen as an important part of many computing tasks e.g. parsing of a program. Probably the most important thing that Snobol demonstrated was the power of pattern matching in a programming language, e.g. it is possible to define a pattern for a title that would include Mr, Mrs, Miss, Rev, etc and search for this pattern in a text using Snobol. Like Lisp it is generally available as an interpreter rather than a compiler, but compiled versions do exist, and are often called Spitbol. Pattern matching capabilities are now to be found in many editors and this makes them very powerful and useful tools. It is in the area of text manipulation that Snobol's greatest contribution to program language development lies.

6.8   Second Generation Languages

6.8.1   PL/1 and Algol 68

It is probably true that Fortran, Algol 60 and Cobol are the three main first generation high level languages. The 60's saw the emergence of PL/1 and Algol 68. PL/1 was a synthesis of features of Fortran, Algol 60 and Cobol. It was soon realised that whilst PL/1 had great richness and power of expression this was in some ways offset by the greater difficulties involved in language definition and use.

These latter problems were true of Algol 68 also. The report introduced its own syntactic and semantic conventions and thus forced upon the prospective user another stage in the learning process. However it has a small but very committed user population who like the very rich facilities provided by the language.

6.8.2 Simula

Another strand that makes up program language development is provided by Simula. It is a general purpose programming language developed by Dahl, Myhrhaug and Nygaard of the Norwegian Computing Centre. The most important contribution that Simula makes is the provision of language constructs that aid the programming of complex, highly interactive problems. It is thus heavily used in the areas of simulation and modelling. It was effectively the first language to offer the opportunity of object orientated programming, and we will back to this very important development in programming languages later in this chapter.

6.8.3 Pascal

The designer of Pascal, Niklaus Wirth, had participated in the early stages of the design of Algol 68 but considered that the generality and complexity of Algol 68 was a move in the wrong direction. Pascal like Algol 68 had its roots in Algol 60 but aimed at providing expressive power through a small set of straightforward concepts. This set is relatively easy to learn and helps in producing readable and hence more comprehensible programs.

It became the language of first choice within the field of computer science during the 1970s and 1980s, and the comment by Wirth sums up the language very well .. *although Pascal had no support from industry, professional societies, or government agencies, it became widely used. The important reason for this success was that many people capable of recognising its potential actively engaged themselves in its promotion. As crucial as the existence of good implementations is the availability of documentation. The conciseness of the original report made it attractive for many teachers to expand it into valuable textbooks. Innumerable books appeared between 1977 and 1985, effectively promoting Pascal to become the most widespread language used in introductory programming courses. Good course material and implementations are the indispensable prerequisites for such an evolution. ..*

6.8.4 APL

APL is another interesting language of the 60's. It was developed by Iverson in the early 1960's and was available by the mid to late 60's. It is an interpretive vector and matrix based language with an extensive set of operators for the manipulation of vectors, arrays etc of whatever data type. As with Algol 68 it has a small but dedicated user population. A possibly unfair comment about APL programs is that you do not debug them, but rewrite them!

6.8.5 Basic

Basic stands for *B*eginners *A*ll Purpose *S*ymbolic *I*nstruction *C*ode, and was developed by Kemeny and Kurtz at Dartmouth during the 1960's. Its name gives a clue to its audience and it is very easy to learn. It is generally interpreted, though compiled versions do exist. It is probably the most heavily used language on micros and home computers. It has proved to be well suited to the rapid development of small programs. It is much criticised because it lacks features that encourage or force the adoption of sound programming techniques.

6.8.6 C

There is a requirement in computing to be able to access directly or at least efficiently the underlying machine. It is therefore not surprising that computer professionals have developed high level languages to do this. This may well seem a contradiction, but it can be done to quite a surprising degree. Some of the earliest published work was that of Martin Richards and the development of BCPL.

This language directly influenced the work of Ken Thompson and can be clearly seen in the programming languages B and C. The UNIX operating system is almost totally written in C and demonstrates very clearly the benefits of the use of high level languages wherever possible.

With the widespread use of UNIX within the academic world C gained considerable ground during the 1970s and 1980s. UNIX systems also offered much to the professional software developer, and became widely used for large scale software development, as as Ritchie says .. *C is quirky, flawed, and an enormous success. While accidents of history surely helped, it evidently satisfied a need for a system implementation language efficient enough to displace assembly language, yet sufficiently abstract and fluent to describe algorithms and interactions in a wide variety of environments.*

6.9 Some Other Strands in Language Development

There are many strands that make up program language development and some of them are introduced here.

6.9.1 Abstraction, Stepwise Refinement and Modules

Abstraction has proved to be very important in programming. It enables a complex task to be broken down into smaller parts concentrating on *what* we want to happen rather than *how* we want it to happen. This leads almost automatically to the ideas of stepwise refinement and modules, with collections of modules to perform specific tasks or steps.

6.9.2 Structured Programming

Structured programming in its narrowest sense concerns itself with the development of programs using a small but sufficient set of statements and in particular control statements. It has had a great effect on program language design and most languages now support the minimal set of control structures.

Introduction to Programming Languages

In a broader sense structured programming subsumes other objectives including simplicity, comprehensibility, verifiability, modifiability and maintenance of programs.

6.9.3 Standardisation

The purposes of a standard are quite varied and include:–

- **Investment in people:** by this we mean that the time spent in learning a standard language pays off in the long term, as what one learns is applicable on any hardware/software platform that has a standard conformant compiler;

- **Portability:** one can take the code one has written for one hardware/software platform and move it to any hardware/software platform that has a standard conformant compiler;

- **Known reference point:** when making comparisons one starts with reference to the standard first, and then between the additional functionality;

and these are some but not all of the reasons for the use of standards. Their importance is summed up beautifully by Ronald G Ross in his introduction to the Cannan and Otten book on the SQL standard, *... Anybody who has ever plugged in an electric cord into a wall outlet can readily appreciate the inestimable benefits of workable standards. Indeed, with respect to electrical power, the very fact that we seldom even think about such access (until something goes wrong) is a sure sign of just how fundamentally important a successful standard can be.*

Appendix A contains notes on what standards apply at this time for the languages covered.

6.10 Ada

Ada represents the culmination of many years of work in program language development. It was a collective effort and the main aim was to produce a language suitable for programming large scale and real time systems. Work started in 1974 with the formulation of a series of documents by the American Department of Defence (DoD), which lead to the Steelman documents. It is a modern algorithmic language with the usual control structures, and facilities for the use of modules and allows separate compilation with type checking across modules.

Ada is a powerful and well engineered language. Its widespread use is certain as it has the backing of the DoD. However it is a large and complex language and consequently requires some effort to learn. It seems unlikely to be widely used except by a small number of computer professionals.

6.11 Modula

Modula was designed by Wirth during the 1970's at ETH, for the programming of embedded real time systems. It has many of the features of Pascal, and can be taken for Pascal at a glance. The key new features that Modula introduced were those of processes and monitors.

As with Pascal it is relatively easy to learn and this makes it much more attractive than Ada for most people, achieving much of the capability, without the complexity.

6.12 Modula 2

Wirth carried on developing his ideas about programming languages and the culmination of this can be seen in Modula 2, and in his words

...In 1977, a research project with the goal to design a computer system (hardware and software) in an integrated approach, was launched at the Institut fur Informatik of ETH Zurich. This system (later to be called Lilith) was to be programmed in a single high level language, which therefore had to satisfy requirements of high level system design as well as those of low level programming of parts that closely interact with the given hardware. Modula 2 emerged from careful design deliberations as a language that includes all aspects of Pascal and extends them with the important module concept and those of multi-programming. Since its syntax was more in line with Modula than Pascal's the chosen name was Modula 2. ...

The language's main additions with regard to Pascal are:

- the module concept, and in particular the facility to split a module into a definition part and an implementation part.
- A more systematic syntax which facilitates the learning process. In particular, every structure starting with a keyword also ends with a keyword, i.e. is properly bracketed.
- The concept of process as the key to multi-programming facilities.
- So called low level facilities which make it possible to breach the rigid type consistency rules and allow to map data with Modula 2 structure onto a store without inherent structure.
- The procedure type which allows procedures to be dynamically assigned to variables.

A sad feature of Modula 2 has been the long time taken to arrive at a standard for the language.

6.13 Other Language Developments

The following is a small selection of language developments that the authors find interesting – they may well not be included in other peoples coverage.

6.13.1 Logo

Logo is a language that was developed by Papert and colleagues at the Artificial Intelligence Laboratory at MIT. Papert is a professor of both mathematics and education, and has been much influenced by the psychologist Piaget. The language is used to create learning environments in which children can communicate with a computer. The language is primarily used to demonstrate and help children develop fundamental concepts of mathematics. Probably the *turtle* and *turtle geometry* are known by educationalists outside of the context of Logo. Turtles have been incorporated into the Smalltalk computer system developed at Xerox Palo Alto Research Centre – Xerox PARC.

6.13.2 Postscript, TeX and LaTeX

The 80's have seen a rapid spread in the use of computers for the production of printed material. The above languages are each used in this area quite extensively.

Postscript is a low level interpretive programming language with good graphics capabilities. Its primary purpose is to enable the easy production of pages containing text, graphical shapes and images. It is rarely seen by most end users of modern desk top publishing systems, but underlies many of these systems. It is supported by an increasing number of laser printers and type-setters.

TeX is a language designed for the production of mathematical texts, and was developed by Donald Knuth. It linearises the production of mathematics using a standard computer keyboard. It is widely used in the scientific community for the production of documents involving mathematical equations.

LaTex is Leslie Lamport's version of TeX, and is regarded by many as more friendly. It is basically a set of macros that hide raw TeX from the end user. The TeX/LaTeX ration is probably 1 to 9 (or so I'm reliably informed by a TeXie).

6.13.3 Prolog

Prolog was originally developed at Marseille by a group lead by Colmerauer in 1972/73. It has since been extended and developed by a variety of people including Pereira (L.M.), Pereira (F), Warren and Kowalski. Prolog is unusual in that it is a vehicle for *logic* programming. Most of the languages described here are basically algorithmic languages and require a specification of *how* you want something done. Logic programming concentrates on the *what* rather than the *how*. The language appears strange at first, but has been taught by Kowalski and others to 10 year old children at schools in London.

6.13.4 SQL

SQL stands for Structured Query Language, and was originally developed by a variety of people mainly working for IBM in the San Jose Research Laboratory. It is a relational database language, and enables programmers to define, manipulate and control data in a relational database. Simplistically a relational database is seen by a user as a collection of tables, comprising rows and columns. It has become the most important language in the whole database field.

6.13.5 ICON

Icon is in the same family as Snobol, and is a high-level general purpose programming language that has most of the features necessary for efficient processing of non-numeric data. Griswold (one of the original design team for Snobol) has learnt much since the design and implementation of Snobol, and the language is a joy to use in most areas of text manipulation.

Available for most systems via anonymous FTP from a number of sites on the Internet.

6.14 Object Orientated Programming – OOP

OOP represents a major advance in program language development. The concepts that this introduces include

- classes
- objects
- messages
- methods

and these in turn draw on the ideas found in more conventional programming languages and correspond to

- extensible data types
- instances of a class
- dynamically bound procedure calls
- procedures of a class

Inheritance is a very powerful high level concept introduced with OOP. It enables an existing data type with its range of valid operations to form the basis for a new class, with additional data types added with corresponding operations, and the new type is compatible with the original.

As was mentioned earlier the first language to offer functionality in this area was Simula, and thus the ideas originate in the 1960s. The book *Simula Begin* by Birtwistle, Dahl, Myhrhaug, Nygaard is well worth a read as it represents one of the first books to introduce the concepts of OOP.

6.14.1 Oberon and Oberon 2

As Wirth says, *The programming language Oberon is the result of a concentrated effort to increase the power of Modula-2 and simultaneously to reduce its complexity. Several features were eliminated, and a few were added in order to increase the expressive power and flexibility of the language.*

Oberon and Oberon 2 are thus developments beyond Modula 2. The main new concept added to Oberon was that of type extension. This enables the construction of new data types based on existing types and to take advantage of what has already been done for that existing type.

Language constructs removed included

- variant records
- opaque types
- enumeration types
- sub-range types

Introduction to Programming Languages

- local modules
- WITH statement
- type transfer functions
- concurrency

The short paper by Wirth provides a fuller coverage. It is available at ETH via anonymous FTP.

6.14.2 Smalltalk

Language plus use of a computer system.

Smalltalk has been under development by the Xerox PARC Learning Research Group since the 1970's. In their words *Smalltalk is a graphical, interactive programming environment. As suggested by the personal computer vision, Smalltalk is designed so that every component in the system is accessible to the user and can be presented in a meaningful way for observation and manipulation. The user interface issues in Smalltalk revolve around the attempt to create a visual language for each object. The preferred hardware system for Smalltalk includes a high resolution graphical display screen and a pointing device such as a graphics pen or mouse. With these devices the user can select information viewed on the screen and invoke messages in order to interact with the information.* Thus Smalltalk represents a very different strand in program language development. The ease of use of a system like this has long been appreciated and was first demonstrated commercially by the Macintosh micro-computers.

Wirth has spent some time at Xerox PARC and has been influenced by their work, and in his own words *the most elating sensation was that after sixteen years of working for computers the computer now seemed to work for me.* This influence can be seen in the design of the Lilith machine, the original Modula 2 engine and in the development of Oberon as both language and operating system.

6.14.3 C++

Stroustrup did his Ph.D thesis at the Computing Laboratory, Cambridge University England, and worked with Simula. He had previously worked with Simula at the University of Aarhus in Denmark. His comments are illuminating... *but was pleasantly surprised by the way the mechanisms of the Simula language became increasingly helpful as the size of the program increased. The class and co-routine mechanisms of Simula and the comprehensive type checking mechanisms ensured that problems and errors did not (as I - and I guess most people - would have expected) grow linearly with the sizeof the program. Instead, the total program acted like a collection of very small (and therefore easy to write, comprehend and debug) programs rather than a single large program.*

He designed C++ to provide Simula's functionality within the framework of C's efficiency, and he has succeeded in this goal. C++ is the most widely used OOP at this time. The major disadvantage now concerns the largely incompatible class libraries that exist. It is hoped that the various standards bodies address this problem in the immediate future.

6.15 Fortran 90

Almost as soon as the Fortran 77 standard was complete and published work began on the next version. The language draws on many of the ideas covered in this chapter and these help to make Fortran 90 a very promising language for the future. Some of the new features include

new source form, with blanks being significant and names being up to 31 characters;

better control structures;

control of the precision of numerical computation;

array processing;

pointers;

user defined data types and operators;

procedures;

modules;

recursion;

dynamic storage allocation;

We will look into all of these in turn.

6.16 The Future and Further Sources

There are already developments in the pipeline. The first of these is the Fortran 1995 standard, and the second are the developments in the High Performance Fortran area.

6.16.1 Fortran 1996

There is work under way for a 1996 standard and details of this can be obtained from a variety of sources on the Internet. Firstly we have a clear up of some of the areas in the standard that have emerged as requiring clarification. Secondly Fortran 96 has added a number of concepts notably:–

- the FORALL construct;
- PURE and ELEMENTAL procedures;

and both of these help considerably in parallelisation of code. We also have structure and pointer default initialisation, which extend the power and usefulness of pointers.

Part two of the standard (ISO/IEC 1539-2:1994) adds the functional specification for varying length character data type, and this extends the usefulness of Fortran for character applications very considerably.

Introduction to Programming Languages

6.16.2 High Performance Fortran – HPF

To quote from the High Performance Fortran Language Specification, Version 1.1, November 10, 1994:–

The High Performance Fortran Forum (HPFF) was founded as a coalition of industrial and academic groups working to suggest a set of standard extensions to Fortran to provide the necessary information. Its intent was to develop extensions to Fortran that provide support for high performance programming on a wide variety of machines, including massively parallel SIMD and MIMD systems and vector processors. From its very beginning, HPFF included most vendors delivering parallel machines, a number of government laboratories and many university research groups. Public input was encouraged to the greatest extent possible.

A number of suppliers now provide HPF Fortran extensions, and this generally based on Fortran 90, rather than Fortran 77.

6.16.3 Network Sources

There is a list called comp-fortran-90 at MAILBASE in the United Kingdom and you can join by sending the following message:–

>   join comp-fortran-90 <first name> <last name>

to MAILBASE@MAILBASE.AC.UK. You will have access to people involved in Fortran 90 standardisation, language implementors for most of the hardware and software platforms, people using Fortran in many very specialised areas, people teaching Fortran etc. The first thing to do after subscription is to obtain the F90 Frequently Asked Questions (FAQ). This can be obtained either by anonymous ftp to mailbase.ac.uk in directory:–

>   /pub/lists/comp-fortran-90/files/

or via Gopher and world wide web browsers and the Universal Resource Locator (URL) is http://mailbase.ac.uk/, or by sending an e-mail message to

>   MAILBASE@MAILBASE.AC.UK

containing the message:–

>   send comp-fortran-90 faq-f90

This prvides a wealth of valuable information regarding Fortran 90. Some other sources are:–

>   ftp.ifremer.fr/ifremer/ditigo/fortran90

>   http://scg.ex.ac.uk/people/eggen/Fortran90/f90-faq.html

>   http://lenti.med.umn.edu/~MWD/F90-FAQ.HTML

>   http://www.ifremer.fr/ditigo/molagnon/fortran90/engfaq.html

>   and the last three are for people with web browsers, e.g. Mosaic, Netscape etc.

The web sites provide links to other sources world wide, and they are a very useful starting point.

There is also a comp.lang.fortran list available via USENET news. This provides access to people worldwide with enormous combined expertise in all aspects of Fortran. Someone will inevitably have encountered your problem or one very much like it and have one or more solutions.

There are many people on the Internet who will make the time to provide you with very valuable advice. As a point of network etiquette please do not waste bandwidth with questions that are answered in the FAQ. Please also spend some time developing an understanding of your problem and making some attempt to see if the answer lies in the documentation or manuals. In computing services and technical support many user problems are labelled RTFM – read the fabulous manual.

6.17 Summary

It is hoped that the reader now has some idea about the wide variety of uses that programming languages are put to.

Fortran 90 now has much more in common with Ada and Modula 2 than one might first have expected from its origins. This is reflected in the UK by the formation under the auspices of the BCS of the Modular Languages Specialist Interest Group, which provides coverage of Ada, Fortran 90, Modula 2, Modula 3, Oberon 2 and Pascal.

6.18 Bibliography

Fortran 90 Standard ISO/IEC 1539:1991

- The definition of the language and the programmers bible. Copies are available for reference in the Computer Centre.

Fortran 96 Draft Standard, ISO/IEC 1539:1996
Varying Length Character Strings in Fortran, ISO/IEC 1539-2:1994.

- The draft is currently available via anonymous FTP from a variety of sources. Consult the Fortran 90 FAQ for up to date details, or visit the Fortran Market!

High Performance Fortran Language Specification, HPFF, Version 1.1, November 1994.

- This is available via anonymous FTP from a variety of sources. Consult the Fortran 90 FAQ for up to date details, or visit the Fortran Market!

Adobe Systems Incorporated, *Postscript Language: Tutorial and Cookbook*, Addison Wesley.

Adobe Systems Incorporated, *Postscript Language: Reference Manual,* Addison Wesley.

Adobe System Incorporated, *Postscript Language: Program Design,* Addison Wesley.

- The three books provide a comprehensive coverage of the facilities and capabilities of Postscript.

ACM SIG PLAN, *History of Programming Languages Conference – HOPL-II*, ACM Press.

- One of the best sources of information on programming language developments, from an historical perspective. The is coverage of Ada, Algol 68, C, C++, CLU, Concurrent Pascal, Formac, Forth, Icon, Lisp, Pascal, Prolog, Smalltalk and Simulation Languages by the people involved in the original design and or implementation. Very highly recommended. This is the second in the HOPL series, and the first was edited by Wexelblat. Details are given later.

Adams, Brainerd, Martin, Smith, Wagener, *Fortran 90 Handbook: Complete ANSI/ISO Reference*, McGraw Hill.

- A complete coverage of the language. As with the Metcalf and Reid book some of the authors were on the X3J3 committee. Expensive, but very thorough.

Annals of the History of Computing, *Special Issue: Fortran's 25 Anniversary*, ACM publication.

- Very interesting comments, some anecdotal, about the early work on Fortran.

Birtwistle G.M., Dahl O. J., Myhrhaug B., Nygaard K., *SIMULA BEGIN*, Chartwell-Bratt Ltd.

- A number of chapters in the book will be of interest to programmers unfamiliar with some of the ideas involved in a variety of areas including systems and models, simulation, and co-routines. Also has some sound practical advice on problem solving.

Brinch-Hansen P., *The Programming Language Concurrent Pascal*, IEEE Transactions on Software Engineering, June 1975, 199-207.

- Looks at the extensions to Pascal necessary to support concurrent processes.

Cannan S., Otten G., *SQL – The Standard Handbook*, McGraw Hill.

- Very thorough coverage of the SQL standard, ISO 9075:1992(E).

Chivers I. D. and Clark M. W., *History and Future of Fortran*, Data Processing, vol. 27 no 1, January/February 1985.

- Short article on an early draft of the standard, around version 90.

Date C. J., *A Guide to the SQL Standard*, Addison Wesley.

- Date has written extensively on the whole database field, and this book looks at the SQL language itself. As with many of Dates works quite easy to read. Appendix F provides a useful SQL bibliography.

Geissman L. B., *Separate Compilation in Modula2 and the structure of the Modula2 Compiler on the Personal Computer Lilith*, Dissertation 7286, ETH Zurich

Jacobi C., *Code Generation and the Lilith Architecture*, Dissertation 7195, ETH Zurich

- Fascinating background reading concerning Modula2 and the Lilith architecture.

Goldberg A., and Robson D., *Smalltalk 80: The language and its implementation*, Addison Wesley.

- Written by some of the Xerox PARC people who have been involved with the development of Smalltalk. Provides a good introduction (if that is possible with the written word) of the capabilities of Smalltalk.

Goos and Hartmanis (Eds), *The Programming Language Ada - Reference Manual*, Springer Verlag.

- The definition of the language.

Griswold R. E., Poage J. F., Polonsky I. P., *The Snobol4 Programming Language*, Prentice-Hall.

- The original book on the language. Also provides some short historical material on the language.

Griswold R. E., Griswold M. T., *The Icon Programming Language*, Prentice-Hall.

- The definition of the language with a lot of good examples. Also contains information on how to obtain public domain versions of the language for a variety of machines and operating systems.

Hoare C.A.R., *Hints on Programming Language Design*, SIGACT/SIGPLAN Symposium on Principles of Programming Languages, October 1973.

- The first sentence of the introduction sums it up beautifully: *I would like in this paper to present a philosophy of the design and evaluation of programming languages which I have adopted and developed over a number of years, namely that the primary purpose of a programming language is to help the programmer in the practice of his art.*

Jenson K., Wirth N., *Pascal: User Manual and Report*, Springer Verlag.

- The original definition of the Pascal language. Understandably dated when one looks at more recent expositions on programming in Pascal.

Kemeny J.G., Kurtz T.E., *Basic Programming*, Wiley.

- The original book on Basic by its designers.

Kernighan B. W., Ritchie D. M., *The C Programming Language,* Prentice Hall: Englewood Cliffs, New Jersey.

- The original work on the C language, and thus essential for serious work with C.

Kowalski R., *Logic Programming in the Fifth Generation*, The Knowledge Engineering Review, The BCS Specialist Group on Expert Systems.

- A short paper providing a good background to Prolog and logic programming, with an extensive bibliography.

Knuth D. E., *The TeXbook*, Addison Wesley.

- Knuth writes with an tremendous enthusiasm and perhaps this is understandable as he did design TeX. Has to be read from cover to cover for a full understanding of the capability of TeX.

Lyons J., *Chomsky*, Fontana/Collins, 1982.

- A good introduction to the work of Chomsky, with the added benefit that Chomsky himself read and commented on it for Lyons. Very readable.

Malpas J., *Prolog: A Relational Language and its Applications*, Prentice-Hall.

- A good introduction to Prolog for people with some programming background. Good bibliography. Looks at a variety of versions of Prolog.

Marcus C., *Prolog Programming: Applications for Database Systems, Expert Systems and Natural Language Systems*, Addison Wesley.

- Coverage of the use of Prolog in the above areas. As with the previous book aimed mainly at programmers, and hence not suitable as an introduction to Prolog as only two chapters are devoted to introducing Prolog.

Metcalf M. and Reid J., *Fortran 90 Explained*, Oxford Science Publications, OUP.

- A clear compact coverage of the main features of Fortran 8x. Reid was secretary of the X3J3 committee.

Mossenbeck H., *Object-Orientated Programming in Oberon-2*, Springer–Verlag.

- One of the best introductions to OOP. Uses Oberon-2 as the implementation language. Highly recommended.

Papert S., *Mindstorms - Children, Computers and Powerful Ideas*, Harvester Press

- Very personal vision of the uses of computers by children. It challenges many conventional ideas in this area.

Sammett J., *Programming Languages: History and Fundamentals*, Prentice Hall.

- Possibly the most comprehensive introduction to the history of program language development – ends unfortunately before the 1980's.

Sethi R., *Programming Languages: Concepts and Constructs*, Addison Wesley.

- Eminently readable and thorough coverage of programming languages. The annotated bibliographic notes at the end of each chapter and the extensive bibliography make it a very useful book.

Reiser M., Wirth N., *Programming in Oberon – Steps Beyond Pascal and Modula*, Addison Wesley.

- Good introduction to Oberon. Revealing history of the developments behind Oberon.

Reiser M., *The Oberon System: User Guide and Programmer's Manual*, Addsion Wesley.

- How to use the Oberon system, rather than the language.

Young S. J., *An Introduction to Ada, 2nd Edition*, Ellis Horwood.

- A readable introduction to Ada. Greater clarity than the first edition.

Wexelblat, *History of Programming Languages, HOPL I*, ACM Monograph Series, Academic Press.

- Very thorough coverage of the development of programming languages up to June 1978. Sessions on Fortran, Algol, Lisp, Cobol, APT, Jovial, GPSS, Simula, JOSS, Basic, PL/I, Snobol and APL, with speakers involved in the original languages. Very highly recommended.

Wirth N., *An Assessment of the Programming Language Pascal*, IEEE Transactions on Software Engineering, June 1975, 192-198.

Wirth N., *History and Goals of Modula2*, Byte, August 1984, 145-152.

- Straight from the horse's mouth!

Wirth N., *On the Design of Programming Languages*, Proc. IFIP Congress 74, 386-393, North Holland, Amsterdam.

Wirth N., *The Programming Language Pascal*, Acta Informatica 1, 35-63, 1971.

Wirth N., *Modula: a language for modular multi- programming*, Software Practice and Experience, 7, 3-35, 1977.

Wirth N., *Programming in Modula2*, Springer Verlag.

- The original definition of the language. Essential reading for anyone considering programming in Modula2 on a long term basis.

Wirth N. *Type Extensions*, ACM Trans. on Prog. Languages and Systems, 10, 2 (April 1988), 2004-214

Wirth N. *From Modula 2 to Oberon*, Software – Practice and Experience, 18,7 (July 1988), 661-670

Wirth N., Gutknecht J., *Project Oberon: The Design of an Operating System and Compiler*, Addison Wesley.

- Fascinating background to the development of Oberon. Highly recommended for anyone involved in large scale program development, not only in the areas of programming languages and operating systems, but more generally.

7
Introduction to Programming

'Though this be madness, yet there is method in't'
Shakespeare.

'Plenty of practice' he went on repeating, all the time that Alice was getting him on his feet again. 'plenty of practice.'
The White Knight, *Through the Looking Glass and What Alice Found There,* Lewis Carroll.

Aims

The aims of the chapter are:–

- to introduce the idea that there is a wide class of problems that may be solved with a computer, and that there is a relationship between the kind of problem to be solved and the choice of programming language that is used to solve that problem;
- to give some of the reasons for the choice of Fortran 90;
- to introduce the fundamental components or kinds of statements to be found in a general purpose programming language;
- to introduce the three concepts of name, type and value;
- to illustrate the above with sample programs based on three of the five intrinsic data types:–
  - character
  - integer
  - real
- to introduce some of the formal syntactical rules of Fortran

7 Introduction to Programming

We have seen that an algorithm is a sequence of steps that will solve a part or the whole of a problem. A program is the realisation of an algorithm in a programming language, and there are at first sight a surprisingly large number of programming languages. The reason for this is that there are a wide range of problems that are solved using a computer, e.g. the telephone company generating itemised bills or the meteorological centre producing a weather forecast. These two problems make different demands on a programming language, and it is unlikely that the same language would be used to solve both.

The range of problems that you want to solve will therefore have a strong influence on the programming language that you use. FORTRAN stands for FORmula TRANslation, which gives a hint of the expected range of problems.

Fortran has been the subject of three standardisation efforts. The first was in 1966, and established Fortran as the major language for numerical programming.

The next standard in 1977 (actually 1978, and thus out by one – a very common programming error, more of this later!) added character handling, but little else in the way of major new features, really tidying up some of the deficiencies of the 1966 standard.

The 1990 standard has altered the language considerably. Some of the major improvements are:–

- source form

    Free source form, names can be up to 31 characters in length, blanks are significant, lines up to 132 characters in length (can be continued up to 39 lines using the continuation character &), ; as statement separator for multiple statements per line, ! as comment symbol; include option for source text from files;

    This offers much that is taken for granted within the Algol family of languages.

- control structures

    The 77 standard offered little in the way of accepted control structures. Fortran 90 has a modern DO statement, with CYCLE and EXIT options and the control part of the DO can be conventional iteration, WHILE or no control clause. There is also a CASE statement.

- numeric precision

    For anyone involved in moving between different hardware platforms and in some instances simply between different compilers on the same hardware platform the specification of precision has always been a problem. This has been quite cleanly addressed in the new standard.

- array processing

    It is now possible to treat arrays as whole objects and simply write statements like

    A = B + C

Introduction to Programming   49

where A, B and C are arrays. Array slicing has been introduced, arrays may be allocated dynamically, and there is a large new set of intrinsic functions to process arrays.

- dynamic behaviour

   Fortran now supports recursion, pointers and allocate/deallocate. This makes algorithms much clearer and understandable, and considerably increases the range of problems that can be easily solved in Fortran.

- user defined data types and operators

   the Algol family has had user defined data types for a long time and there have been many extra hours spent coding round the lack of these facilities in Fortran, prior the 90 standard. This has now been added, with the ability to define the action of operators on these types.

- procedures

   The power of functions and subroutines has been extended greatly with the addition of extra features in this area. Functions can be used to extend existing operators, or add new ones; procedure argument is optional, and keywords can be used; generic procedures can be created;

- modules

   Modules can declare global data, and this is a big improvement over the use of common blocks. A module can make a type definition available; information hiding has been added; interface blocks have been introduced and their use in conjunction with modules aids considerably in the construction of secure large scale programs;

Some of the reasons for choosing Fortran are:–

- it is a modern and expressive language, with much of the power of ADA without the complexity;

- the language is now suitable for a very wide class of both numeric and non-numeric problems;

- the language is widely available in both the educational and scientific sectors;

- a lot of software already exists, written in either Fortran 77 or its, predecessor, Fortran 66. Fortran 66 is also known as Fortran IV. This code can be recompiled with standard conforming Fortran 90 compilers. This thus protects the major investment there is in existing code. Some 15% of code worldwide is estimated to be in Fortran.

There are a few warts however. Given that there has to be backwards compatibility with Fortran 77 some of the syntax is clumsy to say the least. However a considerable range of problems can now be addressed quite cleanly, if one sticks to a subset of the language and adopts a consistent style.

7.1 Elements of a programming language

As with ordinary (so-called *natural*) languages, e.g. English, French, Gaelic German etc., programming languages have rules of syntax, grammar and spelling. The application of the rules of syntax, grammar and spelling in a programming language are more strict. A program has to be unambiguous, since it is a *precise* statement of the actions to be taken. Many everyday activities are rather vaguely defined — *Buy some bread on your way home* — but we are generally sufficiently adaptable to cope with the variations which occur as a result. If, in a program to calculate wages, we had an instruction *Deduct some money* for tax and insurance we could have an awkward problem when the program calculated completely different wages for the same person for the same amount of work every time it was run. One of the implications of the strict syntax of a programming language for the novice is that apparently silly error messages will appear when first writing programs. As with many other *new* subjects you will have to learn some of the jargon to understand these messages.

Programming languages are made up of statements. These statements fall into the following broad categories:–

- **Data description statements**

    These are necessary to describe what kinds of data are to be processed. In the wages program for example, there is obviously a difference between peoples names and the amount of money they earn, i.e. these two things are not the same, and it would not make any sense adding your name to your wages. The technical term for this is *data type;* a wage would be of a different data type (a number) to a surname (a sequence of characters).

- **Control structures**

    A program can be regarded as a sequence of statements to solve a particular problem, and it is common to find that this sequence needs to be varied in practice. Consider again the wages program. It will need to select between a variety of circumstances (say married or single, paid weekly or monthly etc), and also to repeat the program for everybody employed. So there is the need in a programming language for statements to vary and/or repeat a sequence of statements.

- **Data processing statements**

    It is necessary in a programming language to be able to process data. The kind of processing required will depend on the kind or type of data. In the wages program, for example, you will need to distinguish between names and wages. Therefore there must be different kinds of statements to manipulate the different types of data, i.e. *wages* and *names*.

- **Input and output (i/o) statements**

    For flexibility, programs are generally written so that the data that they work on exists *outside* the program. In the wages example the details for each person employed would exist in a *file* somewhere, and there would be a *record* for each person in this file. This means that the program would not have to be modified each time a person left, was ill etc., although the individual records might be updated. It is easier to modify data than to modify a program, and less likely to produce unexpected results. To be able to vary the action there must be some mechanism in a programming language for getting the data *into* and *out of* the program. This is done using input and output statements, sometimes shortened to i/o statements.

Let us now consider a simple program which will read in somebody's name and print it out.

```
PROGRAM Chapter7_Ex01
!
! This program reads in and prints out a name
!
IMPLICIT NONE
CHARACTER (20) :: Name
!
   PRINT *,' Type in your first name, up to 20 characters'
   READ *,Name
   PRINT *,Name
!
END PROGRAM Chapter7_Ex01
```

There are several very important points to be covered here, and they will be taken in turn:–

- Each line is a statement.

- There is a sequence to the statements. The statements will be processed in the order that they are presented, so in this example the sequence is *print, read, print.*

- The first statement names the program. It makes sense to choose a name that conveys something about the purpose of the program.

- The next three lines are *comment* statements. They are identified by a !. Comments are inserted in a program to explain the purpose of the program. They should be regarded as an integral part of all programs. It is essential to get into the habit of inserting comments into your programs straight away.

- The IMPLICIT NONE statement means that there has to be explicit typing of each and every data item used in the program. It is good programming practice to include this statement in every program that you write, as it will trap many errors, some often very subtle in their effect. Using an analogy with a play where there is always a list of the persona involved before the main text of the play, this statement serves the same purpose.

- The CHARACTER (20) statement is a *type* declaration. It was mentioned earlier that there are different kinds of data. There must be some way of telling the programming language that this data is of a certain type, and that therefore certain kinds of operation are allowed and others are banned or just plain stupid! It would not make sense to add a name to a number, e.g. what does Fred+10 mean? So this statement defines that the *variable* NAME to be of type CHARACTER and only character operations are permitted. The concept of a variable is covered in the next section. Character variables of this type can hold up to 20 characters.

- The PRINT statements print out an informative message to the terminal – in this case a guide as to what to type in. The use of informative messages like this throughout your programs is strongly recommended.

- The READ statement is one of the i/o statements. It is an instruction to *read* from the terminal or keyboard; whatever is typed in from the terminal will end up being associated with the variable NAME. Input/output statements will be explained in greater detail in later sections.

- The PRINT statement is another i/o statement. This statement will print out what is associated with the variable NAME and in this case, what you typed in.

- The END PROGRAM statement terminates this program. It can be thought of as being similar to a full stop in natural language, in that it finishes the program, in the same way that a . ends a sentence. Note the use of the name given in the PROGRAM statement at the start of the program.

- Note also the use of the '*' in three different contexts.

- Indentation has been used to make the structure of the program easier to determine. Programs have to be read by human beings and we will look at this in more depth later.

- Lastly, when you do run this program, character input will terminate with the first blank character

The above program illustrates the use of some of the statements in the Fortran language. Let us consider the action of the READ * statement in more detail. In particular, what is meant by a variable and a value.

7.2 Variables — name, type and value

The idea of a variable is one that you are likely to have met before, probably in a mathematical context. Consider the following

    tax payable = gross wages – tax allowances * tax rate

This is a simplified equation for the calculation of wage deductions. Each of the *variables* on the right hand side takes on *a value* for each person, which allows the calculation of the deductions for that person. The above equation expressed in Fortran would be an example of an *arithmetic assignment statement*. There is some arithmetic calculation taking place which yields a value, and this value is then assigned to the variable on the left hand side. This could be expressed in Fortran as

    Tax_Payable = Gross_Wages - Tax_Allowances * Tax_Rate

Note the use of the underscore character and capitalisation to make the statement (and hence program) understandable. Remember a program has to be read and understood by human beings.

The following arithmetic assignment statement illustrates clearly the concepts of name and value, and the difference between the = in mathematics and computing:–

Introduction to Programming

    I=I+1

In Fortran this reads as take the current value of the variable *I* and add one to it, store the new value back into the variable *I*, i.e. *I* takes the value *I+1*. Algebraically,

    i=i+1

does not make any sense.

Variables can be of different types, and the following show some of those available in Fortran.

Variable name	Data type	Value stored
Gross_Pay	REAL	9440.30
Tax_Allowances	INTEGER	2042
Tax_Rate	REAL	.7
Name	CHARACTER	ARTHUR

The concept of data type seems a little strange at first, especially as we commonly think of integers and reals as numbers. However, the benefits to be gained from this distinction are considerable. This will become apparent after writing several programs.

Let us consider another program now. This program reads in three numbers, adds them up and prints out both the total and the average.

```
PROGRAM Sum_and_Average
!
! This program reads in three numbers and sums and averages them
!
IMPLICIT NONE
REAL :: N1,N2,N3,Average = 0.0, Total = 0.0
INTEGER :: N = 3
   PRINT *,'Type in three numbers, separated by spaces or commas'
   READ *,N1,N2,N3
   Total= N1+N2+N3
   Average=Total/N
   PRINT *,'Total of numbers is ',Total
   PRINT *,'Average of the numbers is ',Average
END PROGRAM Sum_and_Average
```

7.3 Notes

The program has been given a name that means something.

There are comments at the start of the program describing what the program does.

The IMPLICIT NONE statement ensures that all data items introduced have to occur in a type declaration.

The next two statements are type declarations. They define the variables to be of *real* or *integer* type. Remember integers are *whole* numbers, while real numbers are those which have a *decimal point*. For example, 2 is an integer, while 2.7, 2.00000001, and 2.0 are all real numbers. One of the fundamental distinctions in Fortran is between integers and reals. Type declarations must always come at the start of a program, before any *processing* is done. Note that the variables have been given sensible names, to aid in making the program easier to understand.

The variables Average, Total and N are also given initial values within the type declaration. Variables are initially undefined in Fortran.

The first PRINT statement makes a text message, (in this case what is between the apostrophes) appear at the terminal. As was stated earlier it is good practice to put out a message like this so that you have some idea of what you are supposed to type in.

The READ statement looks at the input from the keyboard (i.e. what you type) and in this instance associates these values with the three variables. These values can be separated by commas (,), spaces ( ), or even by pressing the carriage return key, i.e. they can appear on separate lines.

The next statement actually does some data processing. It adds up the *values* of the three variables (N1, N2, and N3), and assigns the result to the variable Total. This statement is called an *arithmetic assignment statement*, and is covered more fully in the next chapter.

The next statement is another data processing statement. It calculates the average of the numbers entered and assigns the result to Average. We could have actually used the value 3 here instead, i.e. written Average=Total/3 and had exactly the same effect. This would also have avoided the type declaration for N. However the original example follows established programming practice of declaring all variables and establishing their meaning unambiguously. We will see further examples of this type throughout the book.

Indentation has been used to make the structure of the program easier to determine.

The sum and average are printed out with suitable captions or headings. **Do not** write programs without putting captions on the results. It is too easy to make mistakes when you do this, or even forget what each number means.

Finally we have the end of the program and again we have the use of the name in the PROGRAM statement.

Introduction to Programming

7.4 Some more Fortran rules

There are certain things to learn about Fortran which have little immediate meaning, and some which have no logical justification at all, other than historical precedence. Why is a cat called a cat? At the end of some of the chapters there will be a brief summary of these *rules* or regulations when necessary. Here are a few:-

- Source is free format

- Multiple statements may appear on one line and are separated by the semicolon character ';'

- There is an order to the statements in Fortran. Within the context of what you have covered so far, the order is:-

    - PROGRAM statement

    - Type declarations, e.g. IMPLICIT, INTEGER, REAL or CHARACTER

    - Processing and i/o statements

    - END PROGRAM statement

- Comments may appear anywhere in the program, after PROGRAM and before END, and are introduced with a ! character, and can be in line.

- Names may be up to 31 characters in length and include the underscore character

- Lines may be up to 132 characters

- Up to 39 continuation lines are allowed (using the & as continuation character)

- The syntax of the READ and PRINT statement introduced in these examples is

    - READ *format, input-item-list*

    and

    - PRINT *format, output-item-list*

    where *format* is in the examples *, and called list directed formatting and

- *input-item-list* is a list of variable names separated by commas

- *output-item-list* is a list of variables names and or strings enclosed in either ' or " , again separated by commas.

- If the IMPLICIT NONE statement is not used variables that are not explicitly declared will default to REAL if the first letter of the variable name is A-H or O-Z, and to INTEGER if the first letter of the variable name is I-N.

7.5 Good Programming Guidelines

The following are guidelines, and do not form part of the Fortran 90 language definition.

- Use comments to clarify the purpose of both sections of the program and the whole program.
- Chose meaningful names in your programs.
- Use indentation to highlight the structure of the program. Remember that the program has to be read and understood by both humans and a computer.
- Use IMPLICIT NONE in all programs you write to minimise errors.
- Do not rely on the rules for explicit typing, as this is a major source of errors in programming.

7.6 Fortran Character set

Table 1 summarises the Fortran character set.

A–Z	(left brackets (parenthesis)	% percent
0–9	) right brackets (parenthesis)	& ampersand
_ underscore		; semicolon
blank	, comma	< less than
= equal	. decimal point or period	> greater than
+ plus	' apostrophe	? question mark
– minus	: colon	$ currency symbol
* asterisk	! exclamation mark	
/ slash or oblique	" quotation mark	

Table 1

7.6.1 Notes

Lowercase letters are permitted, but not required to be recognised.

7.7 Problems

1. Write a program that will read in your name and address and print them out in reverse order.

How many lines are there in your name and address? What is the maximum number of characters in the longest line in your name and address? What happens at the first blank character of each input line? What characters can be used in Fortran to enclose each line of text typed in and hence not stop at the first blank character? If you use one of the two special characters to enclose text what happens if you start on one line and then press the return key before terminating the text.? The action here will vary between Fortran implementations.

2. Type in the program Sum_And_Average, given in this chapter. Demonstrate that the input may be separated by spaces or commas. Do you need the decimal point? What happens when you type in too much data? What happens when you type in too little?

8
Arithmetic

Taking Three as the subject to reason about —
  A convenient number to state —
We add Seven, and Ten, and then multiply out
  By One Thousand diminished by Eight.
The result we proceed to divide, as you see,
  By Nine Hundred and Ninety and Two:
Then subtract Seventeen, and the answer must be
  Exactly and perfectly true.

Lewis Carroll, *The Hunting of the Snark*

Round numbers are always false.

Samuel Johnson.

Aims

The aims of this chapter are to introduce:–

- the rules for the evaluation of arithmetic expressions to ensure that they are evaluated as you intend;
- the idea of truncation and rounding applied to reals;
- the use of the PARAMETER statement to define or set up constants
- the concepts and ideas involved in numerical computation, e.g.
  - specifying data types using kind type parameters;
  - the concept of numeric models and positional number systems for integer and real arithmetic and their implementation on binary devices;
  - testing the numerical representation of different kind types on a system;

8 Introduction to Arithmetic

Most problems in the education and scientific community require arithmetic evaluation as part of the algorithm. As the rules for the evaluation of arithmetic in Fortran may differ from those that you are probably familiar with, you need to learn the Fortran rules thoroughly. In the previous chapter, we introduced the arithmetic assignment statement, emphasising the concepts of name, type and value. Here we will consider the way that arithmetic expressions are evaluated in Fortran.

The following are the five arithmetic operators available in Fortran:–

Mathematical operation	Fortran symbol or operator
Addition	+
Subtraction	–
Division	/
Multiplication	*
Exponentiation	**

Exponentiation is raising to a power. Note that the exponentiation operator is the * character *twice*.

The following are some examples of valid arithmetic statements in Fortran:–

```
Taxable_Income = Gross_Pay - Deductions
Cost = Bill + Vat + Service
Delta = Deltax/Deltay
Area = Pi * Radius * Radius
Cube = Big ** 3
```

The above expressions are all simple, and there are no problems when it comes to evaluating them. However, now consider the following:–

```
Nett = Gross - Allowances * Tax_Rate
```

This is ambiguous. There is a choice of doing the subtraction before or after the multiplication. Our everyday experience says that the subtraction should take place before the multiplication. However, if this expression was evaluated in Fortran the *multiplication* would be done before the subtraction.

We need to look at three areas here:–

- the rules for forming expressions, the syntax
- the rules for interpreting expressions, the semantics

- and finally the rules for evaluating expressions, optimisation

The syntax rules determine what expressions are valid. The semantics determine a valid interpretation, and once this has been done the compiler can replace this expression with any other mathematically equivalent expression, generally in the interests of optimisation.

Here are the rules for the evaluation of expressions in Fortran:–

- brackets are used to define priority in evaluation of an expression;
- operators have a hierarchy of priority – a precedence. The hierarchy of operators is:–
  - **exponentiation;** when the expression has multiple exponentiation, they are evaluated right to left. For example,

    L=I**J**K

    is evaluated by first raising J to the power K, and then using this result as the exponent for I; more explicitly therefore

    L=I**(J**K)

    Although this is similar to the way in which we might expect an algebraic expression to be evaluated, it is not consistent with the rules for multiplication and division, and may lead to some confusion. When in doubt, use brackets.
  - **multiplication and division**; within successive multiplications and divisions, the rules regarding any mathematically equivalent expression means that to ensure the evaluation you want brackets must be used. For example with

    A=B*C/D*E

    for non-integer numeric types the compiler does not necessarily evaluate in a left to right manner, i.e. evaluate B times C, then divide the result by D and finally take that result and multiply by E. Section 7.2.5 of the standard confuses the issue by stating that within the context of equal precedence that left to right scanning would apply.
  - **addition and subtraction**; as for multiplication and division the rules regarding any equivalent expression apply. However, it is seldom that the order of addition and subtraction is important, unless other operators are involved. Again section 7.2.5 of the standard confuses the issue by stating in the context of equal precedence that left to right scanning applies.

The following are all examples of valid arithmetic expressions in Fortran:–

```
Slope = (Y1-Y2)/(X1-X2)
X1=(-B+((B*B-4*A*C)**0.5))/(2*A)
Q = Mass_D/2*(Mass_A*Veloc_A/Mass_D)**2+((Mass_A * Veloc_A)**2)/2
```

Introduction to Arithmetic

Note that brackets have been used to make the order of evaluation more obvious. It is often possible to write involved expressions without brackets, but, for the sake of clarity, it is often best to leave the brackets in, even to the extent of inserting a few extra ones to ensure that the expression is evaluated correctly. The expression will be evaluated just as quickly with the brackets as without. Also note that none of the expressions are particularly complex. The last one is about as complex as you should try: with more complexity than this it is easy to make a mistake.

The rule regarding any equivalent expression therefore means if A, B and C are numeric that the following are true:–

A + B = B + A

- A + B = B - A

A + B + C = A + (B + C)

and the last is nominally evaluated left to right, as the additions are of equal precedence.

A * B = B * A

A * B * C = A * (B * C)

and again the last is nominally evaluated left to right, as the multiplications are of equal precedence.

A * B - A * C = A * (B - C)

A / B / C = A / (B * C)

The last is true for non-integer numeric types only.

Problems arise when the value that a *faulty* expression yields lies within the range of expected values and the error may well go undetected. This may appear strange at first, but, a computer does exactly what it is instructed. If, through a misunderstanding on the part of a programmer, the program is syntactically correct but logically wrong from the point of view of the problem definition, then this will not be spotted by the compiler. If an expression is complex, break it down into successive statements with elements of the expression on each line, e.g.

```
Temp = B * B - 4 * A * C
X1 = ( - B + ( Temp ** 0.5 ) ) / ( 2 * A )
```

and

```
Moment = Mass_A * Veloc_A
Q = Mass_D / 2 * ( Moment / Mass_D ) **2+( Moment **2) / 2
```

8.1 Rounding and truncation

When arithmetic calculations are performed one of the following can occur:–

- **truncation.** This operation involves throwing away part of the number, e.g. with 14.6 truncating the number to two figures leaves 14.
- **rounding.** Consider 14.6 again. This is rounded to 15. Basically, the number is changed to the nearest whole number. It is still a real number. What do you think will happen with 14.5, will this be rounded up or down?

You must be aware of these two operations. They may occasionally cause problems in division and in expressions with more than one data type.

To see some of the problems that can occur consider the examples below:–

```
REAL    ::  A,B,C
INTEGER ::  I
   A = 1.5
   B = 2.0
   C = A/B
   I = A/B
```

After executing these statements C has the value 0.75, and I has the value zero! This is an example of type conversion across the = sign. The variables on the right are all real, but the last variable on the left is integer. The value is therefore made into an integer by truncation. In this example, 0.75 is real, so I becomes zero when truncation takes place.

Consider now an example where we assign into a real variable (so that no truncation due to the assignment will take place), but where part of the expression on the right hand side involves integer division.

```
INTEGER ::  I,J,K
REAL    ::  Answer
   I=5
   J=2
   K=4
   Answer=I/J*K
```

The value of ANSWER is 8, because the I/J term involves integer division. The expected answer of 10 is not that different from the actual one of 8, and it is cases like this that cause problems for the unwary, i.e. where the calculated result may be close to the actual one. In complicated expressions it would be easy to miss something like this.

To recap, truncation takes place in Fortran

- across an = sign, when a real is assigned to an integer;
- in integer division

Introduction to Arithmetic

It is very important to be careful when attempting *mixed mode arithmetic* – that is, when mixing reals and integers. If a real and integer are together in a division or multiplication, the result of that operation will be real; when addition or subtraction takes place, in a similar situation, the result will also be real. The problem arises when some parts of an expression are calculated using integer arithmetic, and other parts with real arithmetic:–

```
C = A + B - I / J
```

The integer division is carried out before the addition and subtraction, hence the result of I/J is integer, although all the other parts of the expression will be carried out with real arithmetic.

8.2 Example 1: Time taken for light to travel from the Sun to Earth

How long does it take for light to reach the Earth from the Sun? Light travels 9.46 10^{12} Km in one year. We can take a year as being equivalent to 365.25 days. (As all school-children know, the astronomical year is 365 days, 5 hours, 48 minutes and 45.9747 seconds – hardly worth the extra effort.) The distance between the Earth and Sun is about 150,000,000 Km. There is obviously a bit of imprecision involved in these figures, not least since the Earth moves in an elliptical orbit, not a circular one. One last point to note before presenting the program is that the elapsed time will be given in minutes and seconds. Few people readily grasp fractional parts of a year.

```
PROGRAM Time
IMPLICIT NONE
REAL :: Light_Minute, Distance, Elapse
INTEGER :: Minute, Second
REAL , PARAMETER :: Light_Year=9.46*10**12
! Light_year   : Distance travelled by light in one year in Km
! Light_minute : Distance travelled by light in one minute in Km
! Distance   : Distance from sun to earth in Km
! Elapse : Time taken to travel a distance distance in minutes
! Minute : Integer number part of elapse
! Second : Integer number of seconds equivalent to fractional
! part of elapse
!
   Light_minute = Light_Year/(365.25*24.0*60.0)
   Distance = 150.0*10**6
   Elapse = Distance / Light_minute
   Minute = Elapse
   Second = ( Elapse - Minute)*60
   Print *,' Light takes ' , Minute,' Minutes'
   Print *,'                    ' , Second,' Seconds'
   Print *,' To reach the earth from the sun'
END PROGRAM Time
```

The calculation is straightforward; first we calculate the distance travelled by light in one minute, and then use this value to find out how many minutes it takes for light to travel a set distance. Separating the time taken in minutes into whole number minutes and seconds is accomplished by exploiting the way in which Fortran will truncate a real number to an integer on type conversion. The difference between these two values is the part of a minute which needs to be converted to seconds. Given the inaccuracies already inherent in the exercise, there seems little point in giving decimal parts of a second.

It is worth noting that some structure has been attempted by using comment lines to separate parts of the program into fairly distinct chunks. Note also that the comment lines describe the variables used in the program.

Final note, type this program in and compile it. It will generate an error on certain systems. Can you think where the error will occur?

8.3 The PARAMETER statement.

This statement is used to provide a way associating a meaningful name with a *constant* in a program. Consider a program where PI was going to be used a lot. It would be silly to have to type in 3.14159265358 etc. every time. There would be a lot to type and it is likely that a mistake could be made typing in the correct value. It therefore makes sense to set up PI once and then refer to it by name. However, if PI was just a variable then it would be possible to do the following:–

```
REAL :: LI,PI
 .
PI=3.14159265358
 .
PI=4*ALPHA/BETA
```

The PI=4*ALPHA/BETA statement should have been LI=4*ALPHA/BETA. What has happened is that, through a typing mistake (P and L are close together on a keyboard), an error has crept into the program. It will not be spotted by the compiler. Fortran provides a way of helping here with the PARAMETER statement, which should be preceded with a type declaration. The following are correct examples of the PARAMETER statement:–

```
REAL , PARAMETER :: Pi=3.14159265358 , C=2.997925
```

and

```
REAL , PARAMETER :: Charge=1.6021917
```

The advantage of the PARAMETER statement is that you could not then assign another value to Pi, C or Charge. If you tried to do this, the compiler would generate an error message.

Introduction to Arithmetic

A PARAMETER statement may contain an arithmetic expression, so that some relatively simple arithmetic may be performed in setting up these constants. The evaluation must be confined to addition, subtraction, multiplication, division and integer exponentiation. The following examples help to demonstrate the possibilities

```
REAL , PARAMETER :: Parsec = 3.08*10**16 , &
                    Pi = 3.14159265358 , &
                    Radian = 360./Pi
```

8.4 Precision and size of numbers

The precision and the size of a number in computing is directly related to the number of bits allocated to its internal representation. On a large number of computers this is the same as the word size. The following summarises this information for a number of machines.

Machine and word size (bits)	Maximum Integer	Smallest real Largest real
DEC Alpha (64)	(2**63) - 1 9223372036854774807	0.5562684646268004E-308 0.89884656743115785407E+308
DEC VAX (32)	(2**31) - 1 2,147,483,647	0.29E-38 1.7E38
SUN SPARC (32)	(2**31)-1 2,147,483,647	0.29E-38 1.7E38
IBM PC (8, 16 and 32)	(2**31) - 1 2,147,483,647	0.29E-38 1.7E38

The above information is correct at the hardware level, i.e. this is what the underlying hardware supports.

Precision is not the same as accuracy. In this age of digital time-keeping, it is easy to provide an extremely precise answer to the question *What time is it*? This answer need not be accurate, even though it is reported to tenths (or even hundredths!) of a second. Do not be fooled into believing that an answer reported to ten places of decimals must be accurate to ten places of decimals. The computer can only retain a limited precision. When calculations are performed, this limitation will tend to generate inaccuracies in the result. The estimation of such inaccuracies is the domain of the branch of mathematics known as Numerical Analysis.

To give some idea of the problems, consider an imaginary *decimal* computer, which retains two significant digits in its calculations. For example, 1.2, 12.0, 120.0 and 0.12 are all given to two digit precision. Note therefore that 1234.5 would be represented as 1200.0 in this device. When any arithmetic operation is carried out, the result (including any intermediate calculations) will have two significant digits. Thus

    130+12 = 140 (rounding down from 142)

and similarly

    17/3 = 5.7 (rounding up from 5.666666...)

and

    16*16 = 260

Where there are more involved calculations, the results can become even less attractive. Assume we wish to evaluate

    (16*16) / 0.14

We would like an answer in the region of 1828.5718, or, to two significant digits, 1800.0. If we evaluate the terms within the brackets first, the answer is 260/0.14, or 1857.1428; 1900.0 on the two digit machine. Thinking that we could do better, we could re-write the fraction as

    (16/0.14) * 16

This gives a result of 1800.0.

Algebra shows that all these evaluations are equivalent if unlimited precision is available.

Care should also be taken when is one is near the numerical limits of the machine. Consider the following

    Z = B * C / D

where B, C and D are all 10^{30} and we are using a VAX or IBM PC where the maximum real is approximately 10^{38}. Here the product B * C generates a number of 10^{60} – beyond the limits of the machine. This is called *overflow* as the number is too large. Note that we could avoid this problem by retyping this as

    Z = B * (C/D)

where the intermediate result would now be $10^{30}/10^{30}$, i.e. 1.

There is an inverse called *underflow* when the number is too small, which is illustrated below.

    Z = X1 * Y1 * Z1

where X1 and Y1 are 10^{-20} and Z1 is 10^{20}. The intermediate result of X1 * Y1 is 10^{-40} – again beyond the limits of the machine. This problem could have been overcome by retyping as:–

Introduction to Arithmetic

```
Z = X1 * (Y1 * Z1)
```

This is a particular problem for many scientists and engineers with all machines that use 32 bit arithmetic for integer and real calculations. This is because many physical constants, etc are around the limits of the magnitude (large or small) supported by single precision. This is rarely a problem with machines with hardware support for 64 bit arithmetic.

How we get round this problem and how we move our programs from one platform to another making sure that we are working with the same precision and same range of numbers is covered in detail in the next section.

8.5 Health Warning: Optional Reading, Beginners are Advised to Leave until Later

It is very important in scientific programming to know the range and precision of data on the hardware platform on which we are working. The facilities provided in Fortran 90 now allow the programmer to specify the range and precision they wish to use and the compiler will chose an appropriate type.

If it is not possible to offer the precision and range requested the compiler returns an error code. To avoid this happening the programmer needs to query the computer first for details of its data representations, before trying to run a program which specifies range and precision.

In order to do this we use the KIND intrinsic function. (Intrinsic functions are covered in depth in Chapter 14, and Appendix C), e.g.:–

```
REAL :: X
PRINT *,'Kind number for X = ',KIND(X)
```

This will print out the *kind* number used by the your system to represent default REAL variables. These kind numbers are arbitrary and there is usually no meaning attached to them.

Consider the following program which demonstrates the use of the KIND function.

```
PROGRAM Chapter8_01
IMPLICIT NONE
INTEGER       :: i
REAL          :: r
CHARACTER (1) :: c
LOGICAL       :: l
COMPLEX       :: cp
   PRINT *,' Integer  ',KIND(i)
   PRINT *,' Real     ',KIND(r)
   PRINT *,' Char     ',KIND(c)
   PRINT *,' Logical  ',KIND(l)
   PRINT *,' Complex  ',KIND(cp)
END PROGRAM Chapter8_01
```

It is worthwhile actually typing this program in and seeing what answers you get for the system you are working on. Output from a pc compiler is given below:–

```
[FTN90 Version 1.12 Copyright (c)SALFORD SOFTWARE LTD 1992  & ]
[                       (c)THE NUMERICAL ALGORITHMS  GROUP
1991,1992]
      NO ERRORS    [FTN90]
Program entered
   Integer    3
   Real       1
   Char       1
   Logical    2
   Complex    1
```

The following is the output from the DEC Fortran 90 compiler under OPEN VMS on an alpha 2100:–

```
Program entered
   Integer    4
   Real       4
   Char       1
   Logical    4
   Complex    4
```

Thus it is up to each compiler implementation to decide what kind numbers are associated with each type and kind variation. Thus the kind value on its own should not be used across platforms to try and achieve portability.

In fact specifying a kind number specifically is not what is intended by the Fortran 90 standard, so two intrinsic functions:–

   SELECTED_INT_KIND

and

   SELECTED_REAL_KIND

are available instead. They are used to specify the range of numbers for integers and the range and precision of numbers for reals, and the compiler will return the appropriate kind numbers that it has assigned to such representations. These kind numbers can be assigned to parameters which we call *kind type parameters* and they can be used with REAL and INTEGER type declarations. Let's consider the two main numeric types to see how this works.

Introduction to Arithmetic

8.5.1 Selecting different INTEGER Kinds

The Fortran 90 standard specifies that only one **integer** kind needs to be available, but often a machine's architecture or compiler implementation will offer more. Most compiler implementations will offer the following:–

- 8 bit or one byte integers
- 16 bit or two byte integers
- 32 bit or four byte integers

and 64 bit or eight byte integers will be available on certain platforms and implementations, e.g. the DEC alpha. The most common reason for choosing eight bit or sixteen bit integers is to reduce the memory requirements of your program and the most common reason for choosing sixty four bit integers is to solve specialised problems in mathematics requiring large integer numbers.

To choose an integer kind other than the default, you specify the range of the numbers you require it to lie in, in terms of a power of 10, e.g.:–

```
INTEGER, PARAMETER :: First = SELECTED_INT_KIND (2)
INTEGER (First) :: I,J
...
```

selects an integer kind parameter, First, with representation which includes all integers between -10^2 and 10^2, i.e. numbers in the range -100 to 100. The integer kind parameter can be used in brackets after the integer type statement, to specify variables of this integer kind, e.g. I and J.

If there is no integer kind representation for the range specified, the SELECTED_INT_KIND function returns -1. Unfortunately it is not possible to then test for -1 in a type statement, i.e. you will get a compile time error. We suggest that you run the program in section 8.5.9 of this chapter to find the limits of your machine's architecture before trying to specify a kind parameter that it can't support.

8.5.2 Selecting different REAL Kinds

The Fortran 90 standard specifies that there must be at least two representations of the real type, the default plus one other. Often there are more depending on what the underlying hardware can support. When working with real data there are two things to specify, range and precision. The precision is the minimum number of significant digits (all floating point numbers are normalised) to which real numbers are stored, and the range is the power of 10 of the largest number to be represented. So, for example, to specify that a variable R has a kind type that supports 15 significant figures and a range $\pm 10^{-307}$ to $\pm 10^{307}$ we define a real kind parameter, *Long* and then use this with the REAL type declaration for R as follows:–

```
INTEGER, PARAMETER :: Long=SELECTED_REAL_KIND(15,307)
REAL (Long) :: R
```

The only problem is if the underlying hardware can't support this specification, in which case the function will return -1 if the requested precision is unavailable, - 2 if the range is unavailable, and -3 if both are unavailable. As we mentioned earlier with integer kinds, it is not possible to test for negative values in a type declaration, so before trying to use different kind types, and even just the default types, you need to know what kind types your machine supports and the range and precision of them.

8.5.3 Specifying Kind Types for Literal Integer and Real Constants

A literal constant is a data object whose value cannot change. An integer constant 1 is of default integer kind and a real constant 10.3 is a default real constant. If in a program you have chosen a real kind type, other than the default, then to be consistent and also to make sure that all real arithmetic is done to the precision specified, you need to declare all real constants to be of this kind type. This is done by giving the literal constant followed by an underscore and a kind number or kind type parameter, e.g.:–

```
constant_kind
```

For the earlier example with a kind type parameter Long a real literal constant of this type would be given as:–

```
-22.36_Long
```

It is not recommended to use the actual kind number because, as we have seen, these are not portable across machines.

The convention we use throughout the book if we require a numeric kind type other than the defaults is to specify a kind type parameter, e.g:–

```
    INTEGER, PARAMETER :: Long=SELECTED_REAL_KIND(15,307)
```

and then use it with REAL type declarations e.g.

```
    REAL (Long) :: R
```

This still doesn't make programs completely portable across different hardware platforms, so you will firstly need to run a program which tests the range of data representations. Before doing this we need to know a bit more about the underlying representation of numerical data on computer systems.

8.5.4 Positional Number Systems

Most people take arithmetic completely for granted, and rarely think much about the subject. It is necessary to look in a bit more depth at this subject if we are to understand what the computer is doing in this area.

Introduction to Arithmetic

Our way of working with numbers is essentially a positional one. When we look at the number 1024 for example we rarely think of it in terms of 1 * 1000 + 0 * 100 + 2 * 10 + 4 * 1. Thus the normal decimal systems we use in everyday life is a positional one, with a base of 10.

We are probably aware that we can use other number bases, and 2, 8 and 16 are fairly common alternate number bases. As the computer is a binary device it uses base 2.

We are also reasonably familiar with a mantissa exponent or floating point combination when the numbers get very large or very small, e.g. a parsec is commonly expressed as 3.08*10**16, and here the mantissa is 3.08, and the exponent is 10**16.

The above information will help in understanding the way in which integers and reals are represented on computer systems.

8.5.5   Bit Data Type and Representation Model.

The model is only defined for positive integers (or cardinal numbers), where they are represented as a sequence of binary digits, and is based on the model

$$i = \sum_{k=0}^{n-1} b_k 2^k$$

where i is the integer value, n is the number of bits, and b_k is a bit value of 0 or 1, with bit numbering starting at 0, and reading right to left. Thus the integer 43 and bit pattern 101011 is given by

43 = (1*32) + (0*16) + (1*8) + (0*4) + (1*2) + (1*1)

or

43 = (1*2^5) + (0*2^4) + (1*2^3) + (0*2^2) + (1*2^1) + (1*2^0)

8.5.6   Integer Data Type and Representation Model

The integer data type is based on the model

$$i = s \sum_{k=1}^{q} l_k r^{k-1}$$

where i is the integer value, s is the sign, q is the number of digits (always positive), r is the radix or base (integer greater than 1), and l_k is a positive integer (less than r).

A base of 2 is typical so 1023 is

1023 = (1*2^9) + (1*2^8) + (1*2^7) + (1*2^6) + (1*2^5) + (1*2^4) + (1*2^3) + (1*2^2) + (1*2^1) + (1*2^0)

8.5.7 Real Data Type and Representation Model

The real data type is based on the model

$$x = s\, b^e \sum_{k=1}^{m} f_k\, b^{-k}$$

where x is the real number, s is the sign, b is the radix or base (greater than 1), m is the number of bits in the mantissa, e is an integer in the range e_{min} to e_{max} and f_k is a positive number less than b.

This means that with, for example, a 32 bit real there would be 8 bits allocated to the exponent and 24 to the mantissa. One of these bits in each part would be used to represent the sign and is hence called the sign bit. This reduces the numbers of bits that actually can be used to represent the mantissa and exponent to 31 and 7 respectively. There is also the concept of normalisation, where the exponent is adjusted so that the most significant bit is in position 22 – bits are typically numbered 0-22, rather than 1-23. This form of representation is not new, and is documented around 1750 BC, when Babylonian mathematicians used a sexagesimal (radix 60) positional notation. It is interesting that the form they used omitted the exponent!

This is the theoretical basis of the representation of these three data types in Fortran. Remember from chapter two that the computer is essentially a binary device, and works at the lowest level with sequences of zeros and ones.

This information together with the following provides a good basis for writing portable code across a range of hardware.

8.5.8 IEEE 754

For many people there will also be access to floating point arithmetic as defined in IEEE 754. For example all Sun Sparc and Sun x86 based computers use IEEE arithmetic. The DEC F90 compiler also provides an option to allow specification of IEEE floating point arithmetic.

For a description of IEEE 754 the bibliography should be consulted, and is essential reading when working on Sun hardware.

8.5.9 Example 2: Testing the numerical representation of different kind types on a system

You are now ready to write or adapt a program to run on your system in order to test the range of integer kind types, and the range and precision of real kind types.

The following program selects several integer and real kind types and by calling the intrinsic functions KIND, HUGE, PRECISION and EPSILON it produces most of the information you need to know about these kind types.

The functions do the following:

Introduction to Arithmetic

Function	**Action**
KIND(X)	returns kind type number for X.
HUGE(X)	returns largest number in the numerical model representing numbers of the same type kind as X.
PRECISION(X)	returns the decimal precision in the numerical model representing real numbers of the same kind type as X.
EPSILON(X)	returns a positive model number that is almost negligible compared to unity in the model representing numbers of the same kind type as X.

```
PROGRAM C8_ex04
!
! Examples of the use of the kind function and the numeric
inquiry functions
!
! Integer arithmetic
! -----------------
!
! 32 bits is a common word size, and this leads quite cleanly
! to the following
! 8 bit integers           -128     to           127       10**2
! 16 bit integers        -32768     to         32767       10**4
! 32 bit integers   -2147483648 to      2147483647         10**9
!
!
INTEGER                                          :: I
INTEGER ( SELECTED_INT_KIND( 2))         :: I1
INTEGER ( SELECTED_INT_KIND( 4))         :: I2
INTEGER ( SELECTED_INT_KIND( 9))         :: I3

! Real arithmetic
! ---------------
! 32 bits is a common word size, but 64 bits is also available
!
! 32 bit integers 8 bit exponent, 24 bit mantissa
! This applies on both DEC VAX and the Intel family
! of processors, i.e. 80386, 80486.
!
! 64 bits. Two choices here, simply double the precision
! and keep the mantissa the same or alter both.
!
! 64 bit  8/56 exponent/mantissa - same as for 32 bit
```

```
! 64 bit 11/53 exponent/mantissa - now got approximately the same
! precision as for 56 bit mantissa, but the range is now ~ 10**308
! much more useful in the scientific world.

REAL :: R
REAL ( SELECTED_REAL_KIND(6,37)) :: R1
!
! The next has to be commented out on the pc.
!
!REAL ( SELECTED_REAL_KIND(15,38)) :: R2
!
REAL ( SELECTED_REAL_KIND(15,307)) :: R3
   PRINT *,' '
   PRINT *,' Integer values'
   PRINT *,' Kind      Huge'
   PRINT *,' '
   PRINT *,' ',KIND(I  ),' ',HUGE(I  )
   PRINT *,' '
   PRINT *,' ',KIND(I1 ),' ',HUGE(I1 )
   PRINT *,' ',KIND(I2 ),' ',HUGE(I2 )
   PRINT *,' ',KIND(I3 ),' ',HUGE(I3 )
   PRINT *,' '
   PRINT *,' Real values'
   PRINT *,' Kind      Huge                  Precision           epsilon'
   PRINT *,' '
   PRINT *,' ',KIND(R  ),' ',HUGE(R  ),' ',PRECISION(R),' ',EPSILON(R)
   PRINT *,' '
   PRINT *,' ',KIND(R1 ),' ',HUGE(R1 ),' ',PRECISION(R1),' ',EPSILON(R1)
!  PRINT *,' ',KIND(R2 ),' ',HUGE(R2 ),' ',PRECISION(R2),' ',EPSILON(R2)
   PRINT *,' ',KIND(R3 ),' ',HUGE(R3 ),' ',PRECISION(R3),' ',EPSILON(R3)

END PROGRAM C8_ex04
```

and the result of running this program on a pc follows:

Introduction to Arithmetic

```
[FTN90 Version 2.0 Copyright (c)SALFORD SOFTWARE LTD 1992-1994   &
]
[                        (c)THE NUMERICAL ALGORITHMS GROUP
1991,1992,1993]
     NO ERRORS    [FTN90]
Program entered

   Integer values
   Kind      Huge

    3     2147483647

    1     127
    2     32767
    3     2147483647

   Real values
   Kind      Huge                    Precision           epsilon

    1     3.4028235E+38                   6           1.1920929E-07

    1     3.4028235E+38                   6           1.1920929E-07

    2     1.7976931348623157E+308         15          2.2204460492503131E-16
```

Run this program on whatever system you have access to and compare the output.

8.5.10 Example 3: Binary Representation of Different Integer Kind Type Numbers

For those who wish to look at the internal binary representation of integer numbers, with a variety of kinds we have included the following program.

SELECTED_INT_KIND(2) means provide at least an integer representation with numbers between -10^2 and $+10^2$.

SELECTED_INT_KIND(4) means provide at least an integer representation with numbers between -10^4 and $+10^4$.

SELECTED_INT_KIND(9) means provide at least an integer representation with numbers between -10^9 and $+10^9$.

We use the INT function to convert from one integer representation to another.

We use the logical function BTEST to determine whether the binary value at that position within the number is a zero or a one, i.e. if the bit is set.

I_in_Bits is a character string that holds a direct mapping from the internal binary form of the integer and a text string that prints as a sequence of zeros or ones.

```fortran
PROGRAM Integer_Representation
!
! Use the bit functions in Fortran to write out a
! 32 bit integer number as a sequence of zeros and ones
!
INTEGER :: J
INTEGER :: I
INTEGER ( SELECTED_INT_KIND( 2)) :: I1
INTEGER ( SELECTED_INT_KIND( 4)) :: I2
INTEGER ( SELECTED_INT_KIND( 9)) :: I3
CHARACTER (LEN=32) :: I_in_Bits
   PRINT *,' Type in an integer '
   READ * , I
   I1=INT(I,KIND(2))
   I2=INT(I,KIND(4))
   I3=INT(I,KIND(9))
   I_in_Bits=' '
   DO J=0,7
      IF (BTEST(I1,J)) THEN
         I_in_Bits(8-J):8-J)='1'
      ELSE
         I_in_Bits(8-J):8-J)='0'
      END IF
   END DO
   PRINT *,'          1         2         3'
   PRINT *,'1234567890123456789012345678901234567890'
   PRINT *,I1
   PRINT *,I_in_Bits
   DO J=0,15
      IF (BTEST(I2,J)) THEN
         I_in_Bits(15-J:15-J)='1'
      ELSE
         I_in_Bits(15-J:15-J)='0'
      END IF
   END DO
   PRINT *,I2
   PRINT *,I_in_Bits
```

```
      DO J=0,31
        IF (BTEST(I3,J)) THEN
          I_in_Bits(32-J:32-J)='1'
        ELSE
          I_in_Bits(32-J:32-J)='0'
        END IF
      END DO
      PRINT *,I3
      PRINT *,I_in_Bits
END PROGRAM Integer_Representation
```

The DO loop indices follow the convention of an 8 bit quantity starting at bit 0 and ending at bit 7, 16 bit quantities starting at 0 and ending at 15 etc.

The numbers written out follow the conventional mathematical notation of having the least significant quantity at the right hand end of the digit sequence, i.e. with 127 in decimal we have 1 * 100, 2 * 10 and 7 * 1, so 00100001 in binary means 1 * 32 + 1 * 1 decimal.

Try running this program on the system you are using. Does it produce the results you expect? Experiment with a variety of numbers. Try at least the following 0, +1, -1, -128, 127, 128, -32768, 32767, 32768.

8.5.11 Summary of how to select the appropriate KIND type

To write programs that will perform arithmetically in a similar fashion on a variety of hardware therefore requires an understanding of

- the integer data representation model, and in practice the word size of the various integer kind types;
- the real data representation model, and in practice the word size of the various real kind types and the number of bits in both the mantissa and exponent;

Armed with this information we can then choose a kind type that will ensure minimal problems when moving from one platform to another. End of health warning!

8.6 Summary

- learn the rules for the evaluation of arithmetic expressions;
- break expressions down where necessary to ensure that the expressions are evaluated in the way you want;
- take care with truncation due to integer division in an expression. Note that this will only be a problem where both parts of the division are INTEGER;

- take care with truncation due to the assignment statement when there is an integer on the left hand sign of the statement, i.e. assigning a real into an integer variable;

- when you want to set up constants, which will remain unchanged throughout the program, use the PARAMETER statement;

- do not confuse precision and accuracy.

- learn what the default KINDs are for the numeric types you work with, and what the maximum, minimum values and precision for REAL data, and what the maximum and minimum are for INTEGER data.

- you have been introduced to the use of the functions DIGITS, HUGE and PRECISION, and some of the concepts involved in their use. We will look at functions in much greater depth later on.

8.7 Problems

1. Modify the program that read in your name and address to also read in and print out your age, telephone number and sex.

2. One of the easiest ways to write a program is to modify an existing one. The example given earlier, dealing with the time taken for light to travel from the Sun to the Earth, could form the basis of several other programs.

(i) Many communications satellites follow a geosynchronous orbit, some 35,870 Km above the Earths surface. What is the time lag incurred in using one such satellite for a telephone conversation?

(ii) The Moon is about 384,400 Km from the Earth on average. What implications does this have for control of experiments on the Moon? What is the time lag?

(iii) The following table gives the distance in MKm from the Sun to the planets in the Solar System:

Mercury	57.9	Venus	108.2
Earth	149.6	Mars	227.9
Jupiter	778.3	Saturn	1427.0
Uranus	2869.6	Neptune	4496.6
Pluto	5900.0		

Use this information to find the greatest and least time taken to send a message from Earth to the other planets. Assume that all orbits are in the same plane and circular (if it was good enough for Copernicus, its good enough for this example). For all practical purposes, the speed of light in vacuum is a constant, and therefore an excellent candidate for a PARAMETER statement. Use it.

3. Write a program to read in, sum and average five numbers.

4. Write a program to calculate the period of a pendulum. Use the following formula:–

Introduction to Arithmetic

```
T=2*PI*(LENGTH/9.81)**.5
```

Calculate the period for at least 5 values of the length. The length (LENGTH) is in metres, and the time (T) in seconds.

5. Unit pricing: the following table gives the price and weight of various cereals available in the local supermarket.

Cereal	Price	Weight (grams)
Frostys	75	375
Special L	76	250
Rice Crispys	71	295
Rice Crispys	97	440
Bran Bits	85	625
Raisin Bran	84	375
Icicles	67	280
Frostys	58	250
Coco Puffs	77	280
Huffa Puffa Rice	76	230
Friends Oats	74	750
Weetabits	40	375
Welsh Porage Oats	74	750
More	61	250
Korn Flakes	81	500
Korn Flakes	65	375

Which of these gives best value for money, in terms of cost per gram (i.e. unit pricing)? Which gives the poorest value, on the same criterion?

5. Write a program that tests the precision and size of numbers of the system you use. Finding out the word size of the machine you work with is the first step. Then try some multiplication and division, and see what sort of error messages you get as the numbers become too small and too large.

If you have access to more than one Fortran 90 compiler try the program out on the other platforms and look at the differences in error reporting.

8.8 Bibliography

Some understanding of numerical analysis is essential for successful use of Fortran when programming. As Froberg says *numerical analysis is a science – computation is an art*. The following are some of the more accessible books available.

Froberg C.E., *Introduction to Numerical Analysis*, Addison Wesley.

- The short chapter on numerical computation is well worth a read, and it covers some of the problems of conversion between number bases, and some of the errors that are introduced when we compute numerically. The Samuel Johnson quote owes its inclusion to Froberg!

IEEE, *IEEE Standard for Binary Floating-Point Arithmetic, ANSI/IEEE Std 754-1985*, Institute of Electrical and Electronic Engineers Inc.

- The formal definition of IEEE 754.

Knuth D., *Seminumerical Algorithms*, Addison Wesley.

- A more thorough and mathematical coverage than Wakerly. The chapter on positional number systems provides a very comprehensive historical coverage of the subject. As Knuth points out the floating point representation for numbers is very old, and is first documented around 1750 B.C. by Babylonian mathematicians. Very interesting and worthwhile reading.

Sun, *Numerical Computation Guide*, SunPro.

- Very good coverage of the numeric formats for IEEE Standard 754 for Binary Floating-Point Arithmetic. All SunPro compiler products support the features of the IEEE 754 standard.

Wakerly J.F., *Microcomputer Architecture and Programming*, Wiley.

- The chapter on number systems and arithmetic is surprisingly easy. There is a coverage of positional number systems, octal and hexadecimal number system conversions, addition and subtraction of non-decimal numbers, representation of negative numbers, two's complement addition and subtraction, one's complement addition and subtraction, binary multiplication, binary division, bcd or binary coded decimal representation and fixed and floating point representations.
  There is also coverage of a number of specific hardware platforms, including DEC PDP-11, Motorola 68000, Zilog Z8000, TI 9900, Motorola 6809 and Intel 8086. A little old but quite interesting nevertheless.

9
Arrays 1
Some Fundamentals

Thy gifts, thy tables, are within my brain
Full charactered with lasting memory

William Shakespeare, *The Sonnets*.

Here, take this book, and peruse it well:
The iterating of these lines brings gold;

Christopher Marlowe, *The Tragical History of Doctor Faustus*

Aims

The aims of the chapter are to introduce the fundamental concepts of arrays and do loops, in particular:–

- to introduce the ideas of tables of data and some of the formal terms used to describe them:
    - Array
    - Vector
    - List and linear list
- to discuss the array as a random access structure where any element can be accessed as readily as any other:
- to note that the data in an array is all of the same type:
- to introduce the twin concepts of data structure and corresponding control structure:
- to introduce the statements necessary in Fortran to support and manipulate these data structures.

9 Arrays 1: Some Fundamentals

9.1 Tables of data

Consider the following:–

9.1.1 Telephone directory

A telephone directory consists of the following kinds of entries:–

Name	Address	Number
Adcroft A.	61 Connaught Road, Roath, Cardiff	223309
Beale K.	14 Airedale Road, Balham	745 9870
Blunt R.U.	81 Stanlake Road, Shepherds Bush	674 4546
...		
...		
...		
Sims Tony	99 Andover Road, Twickenham	898 7330

This *structure* can be considered in a variety of ways, but perhaps the most common is to regard it as a *table* of data, where there are 3 columns and as many rows as there are entries in the telephone directory.

Consider now the way we extract information from this table. We would scan the name column looking for the name we are interested in, and then read along the row looking for either the address or telephone number, i.e. we are using the name to *look up* the item of interest.

9.1.2 Book catalogue

A catalogue could contain:–

Author(s)	Title	Publisher
Carroll L.	Alice through the looking Glass	Penguin
Knuth D.	Semi-numerical Algorithms	Addison-Wesley
Steinbeck.J	Sweet Thursday	Penguin
Wirth N.	Algorithms plus Data Structures = Programs	Prentice Hall

Arrays 1: Some Fundamentals

Again, this can be regarded as a *table* of data, having 3 columns and many rows. We would follow the same procedure as with the telephone directory to extract the information. We would use the *Name* to *look up* what books are available.

9.1.3 Examination marks or results

This could consist of:–

Name	Physics	Maths	History	Geography	French
Fowler .L.	50	47	89	30	46
Barron.L.W	37	67	65	68	98
Warren.J.	25	45	48	10	36
Mallory.D.	89	56	45	30	65
Codd.S.	68	78	76	98	65

This can again be regarded as a *table* of data. This example has 6 columns and 5 rows. We would again *look up* information by using the Name.

9.1.4 Monthly rainfall

Typically this would consist of:–

Month	Rainfall
January	10.4
February	11.1
March	8.3
April	7.5
May	3.2
June	4.6
July	3.2
August	4.5
September	2.1
October	3.1
November	10.1
December	11.1

In this table there are 2 columns and 12 rows. To find out what the rainfall was in July, we scan the table for July in the Month column and locate the value in the same row, i.e. the rainfall figure for July.

These are just some of the many examples of problems where the data that is being considered has a tabular structure. Most general purpose languages therefore have mechanisms for dealing with this kind of structure. Some of the special names given to these structures include:–

- Linear list
- List
- Vector
- Array

The term used most often here, and in most books on Fortran programming, is *array*.

9.2 Arrays in Fortran

There are three key things to consider here:–

- the ability to refer to a set or group of items by a single name;
- the ability to refer to individual items or members of this set, i.e. look them up;
- the choice of a control structure that allows easy manipulation of this set or array.

9.3 The DIMENSION Attribute

The DIMENSION attribute defines a variable to be an array. This satisfies the first requirement of being able to refer to a set of items by a single name. Some examples are given below:–

```
REAL , DIMENSION(100) :: Wages
INTEGER , DIMENSION(10000) ::   Sample
```

For the variable Wages it is up type REAL and an array of dimension or size 100, i.e. the variable array Wages can hold up to 100 real items.

For the variable Sample it is of type INTEGER and an array of dimension or size 10,000, i.e. the variable Sample can hold up to 10,000 integer items.

9.4 An index

An index enables you to refer to or select individual elements of the array. In the telephone directory, book catalogue, exam marks table and monthly rainfall examples we used the name to index or look up the items of interest. We will give concrete Fortran code for this in the example of monthly rain fall.

Arrays 1: Some Fundamentals

9.5 Control structure

The statement that is generally used to manipulate the elements of an array is the DO statement. It is typical to have several statements controlled by the DO statement, and the block of repeated statements is often called a *DO loop*. Let us look at two complete programs that highlight the above.

9.6 Monthly Rainfall

Let us look at this earlier example in more depth now. Consider the following:–

Month	Associated integer representation	Array and index	Rainfall value
January	1	RainFall(1)	10.4
February	2	RainFall(2)	11.1
March	3	RainFall(3)	8.3
April	4	RainFall(4)	7.5
May	5	RainFall(5)	3.2
June	6	RainFall(6)	4.6
July	7	RainFall(7)	3.2
August	8	RainFall(8)	4.5
September	9	RainFall(9)	2.1
October	10	RainFall(10)	3.1
November	11	RainFall(11)	10.1
December	12	RainFall(12)	11.1

Most of you should be familiar with the idea of the use of an integer as an alternate way of representing a month, e.g. in a date expressed as 1/3/1989, for 1st March 1989 (anglicised style) or 3rd January (americanised style). Fortran, in common with other programming languages, only allows the use of integers as an index into an array. Thus when we write a program to use arrays we have to map between whatever construct we use in everyday life as our index (names in our examples of telephone directory, book catalogue, and exam marks) to an integer representation in Fortran. The following is an example of an assignment statement showing the use of an index.

```
RainFall(1)=10.4
```

We saw earlier that we could use the DIMENSION attribute to indicate that a variable was an array. In the above example Fortran statement our array is called RainFall. In this statement we are assigning the value 10.4 to the first element of the array, i.e. the rainfall for the month of January is 10.4 We use the index 1 to represent the first month. Consider the following statement:–

```
SummerAverage = (RainFall(6) + RainFall(7) + RainFall(8))/3
```

This statement says take the values of the rainfall for June, July and August, add them up and then divide by 3, and assign the result to the variable SummerAverage, thus providing us with the rain fall average for the three summer months – northern hemisphere of course.

9.6.1 Example 1: Rainfall

The following program reads in the 12 monthly values from the terminal, computes the sum and average for the year, and prints the average out.

```
PROGRAM Rain
IMPLICIT NONE
REAL ::    Sum=0.0, Average=0.0
REAL , DIMENSION(12) :: RainFall
INTEGER ::   Month
   PRINT *,' Type in the rainfall values'
   PRINT *,' one per line'
   DO Month=1,12
      READ *, RainFall(Month)
   ENDDO
   DO Month=1,12
      Sum = Sum + RainFall(Month)
   ENDDO
   Average = Sum / 12
   PRINT *,' Average monthly rainfall was'
   PRINT *, Average
END PROGRAM Rain
```

RainFall is the *array name*. The variable Month in brackets is the index. It takes on values from 1 to 12 inclusive, and is used to pick out or select elements of the array. The index is thus a variable and this permits dynamic manipulation of the array at *run time*. The general form of the DO statement is:

```
DO Counter = Start, End, Increment
```

The block of statements that form the loop are contained between the DO statement, which marks the beginning of the block or loop, and the ENDDO statement , which marks the end of the block or loop.

In this program, the DO loops take the form:–

```
DO Month=1,12          start
   ...                        body
ENDDO                  end
```

Arrays 1: Some Fundamentals

The body of the loop in the program above has been indented. This is *not* required by Fortran. However it is good practice and will make programs easier to follow.

The number of times that the DO loop is executed is governed by the last part of the DO statement, i.e. by the:–

```
Counter = Start, End, Increment
```

Start, as it implies, is the initial value which the counter (or index, or control variable) takes. Each time the loop is executed, the value of the counter will be increased by the value of *increment,* until the value of *end* is reached. If *increment* is omitted, it is assumed to be 1. No other element of the DO statement may be omitted. In order to execute the statements within the loop (the *body*) it must be possible to reach *end* from *start.* Thus zero is an illegal value of *increment.* In the event that it is not possible to reach *end,* the loop will not be executed and control will pass to the statement after the end of the loop.

In the example above both loops would be executed 12 times. In both cases, the first time round the loop the variable MONTH would have the value 1, the second time round the loop the variable MONTH would have the value 2 etc., and the last time round the loop MONTH would have the value 12.

9.7 People's Weights

Consider the following:–

Person	Associated integer representation	Array and index	Associated value
Andy	1	Weight(1)	48.7
Barry	2	Weight(2)	76.5
Cathy	3	Weight(3)	58.5
Dawn	4	Weight(4)	65.3
Elaine	5	Weight(5)	88.7
Frank	6	Weight(6)	67.5
Gordon	7	Weight(7)	56.7
Hannah	8	Weight(8)	66.7
Ian	9	Weight(9)	70.6
Jatinda	10	Weight(10)	65.5

We have ten people, with their names as shown. We associate each name with a number — in this case we have ordered the names alphabetically, and the numbers therefore reflect their ordering. WEIGHT is the *array name.* The number in brackets is called the *index.* The index is used to pick out or select elements of the array. Therefore the table is read as 'the first element of the array WEIGHT has the value 48.7, the second element of the array WEIGHT has the value 76.5.

9.7.1 Example 2: People's Weights

There are two examples in the program below:-

```
PROGRAM SumAndAverage
! The program reads up to 10 weights into the array Weight
! Variables used
!     Weight, holds the weight of the people
!     Person, an index into the array
!     Total, total weight
!     Average, average weight of the people
! Parameters used
!     NumberOfPeople ,10 in this case.
! The weights are written out so that they can be checked
!
IMPLICIT NONE
INTEGER , PARAMETER :: Number_Of_People = 10
REAL ::   Total = 0.0, Average = 0.0
INTEGER :: Person
REAL , DIMENSION(Number_of_People) :: Weight
   DO Person=1,Number_Of_People
      READ *,Weight(Person)
      Total = Total + Weight(Person)
   ENDDO
   Average = Total / Number_Of_People
   PRINT *,' Sum of numbers is    ',Total
   PRINT *,' Average Weight is    ',Average
   PRINT *,' 10 Weights were    '
   DO  Person=1,Number_Of_People
      PRINT *,Weight(Person)
   ENDDO
END PROGRAM SumAndAverage
```

9.8 Summary

The DIMENSION attribute declares a variable to be an array. The DIMENSION attribute must come at the start of a program unit, with other *declarative* statements.

The PARAMETER attribute declares a variable to have a fixed value that cannot be changed during the execution of a program. In our example above note that this statement occurs before the other

Arrays 1: Some Fundamentals

declarative statements that depend on it. To recap the statements covered so far, the order is summarised below.

PROGRAM	first statement	
INTEGER REAL CHARACTER	*declarative*	in any order and the DIMENSION and PARAMETER attributes are added here
Arithmetic assignment PRINT * READ * DO ENDDO	*executable*	in any order
END PROGRAM	last statement	

We choose individual members using an index, and these are always of integer type in Fortran.

The DO loop is a very convenient control structure to manipulate arrays, and we use indentation to clearly identify loops.

9.9 Problems

1. Using a DO loop and an array rewrite the program which calculated the average of five numbers (question 3 in chapter 6) and increase the number of values read in from 5 to 10.

2. Write a program to sum and average people's weights.

3. Generalise the above program to allow an arbitrary number of weights to be read in using the Fortran statements we have covered so far. What is a sensible upper bound here? Do you find this a particularly satisfactory solution?

4. Modify the program that read in your name to read in 10 names. Use an array and a DO loop. When you have read the names into the array write them out in reverse order on separate lines.

5. Combine the programs that read in and calculate the average weight with the one that reads in peoples names. The program should read in the weights into one array, and the names into another. Allow 20 characters for the length of a name. Print out a table linking names and weights.

10
Arrays 2
Further Examples

Sir, In your otherwise beautiful poem (The Vision of Sin) there is a verse which reads
Every moment dies a man,
every moment one is born.
Obviously this cannot be true and I suggest that in the next edition you have it read
Every moment dies a man,
every moment 1 1/16 is born.
Even this value is slightly in error but should be sufficiently accurate for poetry.
Charles Babbage in a letter to Lord Tennyson.

Aims

The aims of the chapter are to extend the concepts introduced in the previous chapter and in particular:–

- to introduce the idea of an array with more than one dimension;
- to introduce the corresponding control structure to permit easy manipulation of higher dimensioned arrays;
- to introduce an extended form of the DIMENSION attribute declaration;
- to introduce the corresponding alternative form to the DO statement, to help manipulate the array in this new form;
- to introduce the concept of array element ordering;
- to introduce the DO loop as a mechanism for the control of repetition in general, not just for manipulating arrays;
- to formally define the block DO syntax.

10 Arrays 2: Further Examples

10.1 Higher dimension arrays

There are many instances where it is necessary to have arrays with more than one dimension. Consider the following examples:

10.1.1 Example 1: A Map

Consider the representation of height of an area of land expressed as a two dimensional table of numbers, e.g. we may have some information represented in a simple table as shown below:—

Latitude Longitude	1	2	3	4	5
1	15.1	215.2	250.5	262.6	145.7
2	52.1	73.4	228.1	65.8	243.7
3	56.0	356.3	339.8	113.7	39.4
4	367.8	327.0	66.6	142.4	259.3
5	223.3	66.5	243.3	58.0	28.9

The values in the *array* are the heights above sea level.

A program to manipulate this data structure would involve something like the following:—

```
PROGRAM Locate
! Variables used
! Height - used to hold the heights above sea level
! Longitude - used to represent the longitude
! Latitude - used to represent the latitude
!    both restricted to integer values.
IMPLICIT NONE
INTEGER   :: Latitude , Longitude
REAL , DIMENSION(5,5) :: Height
   DO Latitude = 1,5
      DO Longitude = 1,5
         READ * , Height(Longitude,Latitude)
      ENDDO
   ENDDO
      .
      .
END PROGRAM Locate
```

Note the way in which indentation has been used to highlight the structure in this example. The *inner* loop is said to be *nested* within the outer one. It is very common to encounter problems where nesting is a natural way to express the solution. Nesting is permitted to any depth. Here are an example of a valid nested DO loop.

```
DO                              ! Start of outer loop
   .
   .
    DO                          ! Start of inner loop
       .
       .
    ENDDO                       ! End of inner loop
   .
ENDDO                           ! End of outer loop
```

This example introduces the concept of two indices, and can be thought of as a row and column data structure.

10.1.2 Example 2: Booking arrangements in a theatre or cinema

A theatre or cinema consists of rows and columns of seats. In a large cinema or a typical theatre there would also be more than one level or storey. Thus, a program to represent and manipulate this structure would probably have a two or three dimensional array. Consider the following program extract:–

```
PROGRAM Theatre
IMPLICIT NONE
INTEGER , PARAMETER :: NRows=30, NColumns=30, NFloors=3
INTEGER :: Row,Column,Floor
? , DIMENSION(NRows, NColumns, NFloors) :: Seats
   ...
   DO   Floor=1,NFloors
      DO   Column=1,NCols
         DO   Row=1,NRows
            PRINT *,Seats(Row,Column,Floor)
         ENDDO
      ENDDO
   ENDDO
   ...
END PROGRAM Theatre
```

Note here the use of the term PARAMETER in conjunction with the INTEGER declaration. This is called an entity orientated declaration. An alternative to this is attribute orientated declarations, e.g.

Arrays 2: Further Examples

```
INTEGER   ::  NRows,NColumns,NFloors
PARAMETER ::  NRows=30,NColumns=30,NFloors
```

and we will be using the entity orientated declaration method throughout the rest of the book. This is our recommended method as you only have to look in one place to determine everything that you need to know about an entity.

An interesting question here is what is the best data type for *Seats*. We will leave the choice to you.

10.2 Additional forms of the DIMENSION attribute and DO loop statement

10.2.1 Example 3: Voltage from -20 to +20 volts

Consider the problem of an experiment where the independent variable voltage varies from −20 to +20 volts and the current is measured at 1 volt intervals. Fortran has a mechanism for handling this type of problem:–

```
PROGRAM C10_03
IMPLICIT NONE
REAL , DIMENSION(-20:20) :: Current
REAL :: Resistance
INTEGER :: Voltage
   ...
   DO  Voltage=-20,20
      ...
      Current(Voltage)=Voltage/Resistance
      ...
   ENDDO
   ...
END PROGRAM C10_03
```

We appreciate that, due to experimental error, the voltage will not have exact integer values. However, we are interested in representing and manipulating a set of values, and thus from the point of view of the problem solution and the program this is a reasonable assumption. There are several things to note here:–

- This form of the DIMENSION attribute–

    ```
    DIMENSION(First:Last)
    ```

is of considerable use when the problem has an effective index which does not start at 1 (as implied by the original form of the statement).

- There is a corresponding form of the DO statement which allows processing of problems of this nature. This is shown in the above program. The general form of the DO statement is therefore:–

```
DO counter=start, end, increment
```

where *start, end* and *increment* can be positive or negative. Note that zero is a legitimate value of the dimension limits and of a DO loop index.

10.2.2   Example 4: Longitude from -180 to +180

Consider the problem of the production of a table linking time difference with longitude. The values of longitude will vary from –180 to +180 degrees, and the time will vary from +12 hours to –12 hours. A possible program segment is:–

```
PROGRAM Zone
IMPLICIT NONE
REAL , DIMENSION(-180:180) :: Time
INTEGER :: Degree,Strip
REAL :: Value
   ...
   DO Degree=-180,170,10
      Value=Degree/15.
      DO Strip=0,9
         Time(Degree+Strip)=Value
       ENDDO
   ENDDO
   ...
END PROGRAM Zone
```

10.2.3   Notes

The values of the time are **not** being calculated at every degree interval.

The variable Time is a real variable. It would be possible to arrange for the time to be an integer by expressing it in either minutes or seconds.

This example takes no account of all the wiggly bits separating time zones, or of British Summer Time.

What changes would you make to the program to accommodate +180? What is the time at -180 and +180?

10.3   The DO loop and straight repetition

10.3.1   Example 5: Table of Temperatures

Consider the production of a table of temperatures. The independent variable is the Fahrenheit value; the Celsius temperature is the dependent variable. Strictly speaking, a program to do this does not have to have an array, i.e. the DO loop can be used to control the repetition of a set of statements that make no reference to an array. The following shows a possible program segment.

Arrays 2: Further Examples

```
PROGRAM Convert
IMPLICIT NONE
INTEGER :: Fahrenheit
REAL :: Celsius
   ...
   .
   DO Fahrenheit=-100,200
      ...
      Celsius=(Fahrenheit-32)*5./9.
      PRINT *, Fahrenheit , Celsius
      ...
   ENDDO
   ...
END PROGRAM Convert
```

Note here that the DO statement has been used *only* to control the repetition of a block of statements. In the conversion of Fahrenheit to Celsius, we multiply by 5./9. rather than 5/9; do you recall why?

This is the other use of the DO statement. The DO loop thus has two functions, its use with arrays as a control structure, and its use solely for the repetition of a block of statements.

10.3.2 Example 6: Means and Standard Deviations

In the calculation of the mean and standard deviation of a list of numbers, we may use the following formulae. It is not actually necessary to store the values, nor to accumulate the sum of the values and their squares. In the first case, we would possibly require a large array, while in the second, it is conceivable that the accumulated values (especially of the squares) might be too large for the machine. The following example uses an *updating* technique which avoids these problems, but is still accurate. The DO loop is simply a control structure to ensure that all the values are read in, with the index being used in the calculation of the updates.

```
PROGRAM MeansAndStandardDeviation
! Variables used are
!     Mean  - for the running mean
!     SSQ   - The running corrected sum of squares
!     X     - Input values for which mean and sd required
!     W     - Local work variable
!     SD    - Standard Deviation
!     R     - Another work variable
IMPLICIT NONE
REAL :: Mean=0.0,SSQ=0.0,X,W,SD,R
INTEGER :: I,N
   PRINT *,' ENTER THE NUMBER OF READINGS'
```

```
      READ*,N
      PRINT*,' ENTER THE ',N,' VALUES, ONE PER LINE'
      DO I=1,N
          READ*,X
          W=X-Mean
          R=I-1
          Mean=(R*Mean+X)/I
          SSQ=SSQ+W*W*R/I
      ENDDO
      SD=(SSQ/R)**0.5
      PRINT *,' Mean is ',Mean
      PRINT *,' Standard deviation is ',SD
END PROGRAM MeansAndStandardDeviation
```

10.4 Summary

- Arrays can have up to 7 dimensions.

- DO loops may be nested, but they must not overlap.

- The DIMENSION attribute allows limits to be specified for a block of information which is to be treated in a common way. The limits must be integer, and the second limit must exceed the first e.g.

    ```
    REAL , DIMENSION(-123:-10) :: List
    REAL , DIMENSION(0:100,0:100) :: Surface
    REAL , DIMENSION(1:100) :: Value
    ```

The last example could equally be written

```
REAL , DIMENSION(100) :: Value
```

where the first limit is omitted, and is given the default value 1. The array LIST would contain 114 values, while X would contain 2001.

- A DO statement and its corresponding ENDDO statement define a loop. The DO statement provides a starting value, terminal value, and, optionally, an increment, for its index or counter.

- The increment may be negative, but should never be zero. If it is not present, the default value is 1. It must be possible for the terminating value to be reached from the starting value.

- The counter in a DO loop is ideally suited for indexing an array, but it may be used anywhere that repetition is needed, and of course the index or counter need not be used explicitly.

The formal syntax of the block DO construct is

Arrays 2: Further Examples

```
[ do-construct-name : ] DO [  label ] [ loop-control ]
   [execution-part-construct  ]
[ label ] end-do
```

where the forms of the loop control are

```
[ , ] scalar-variable-name =   scalar-numeric-expression ,
                               scalar-numeric-expression
                               [ , scalar-numeric-expression ]
```

and the forms of the end-do are

```
END DO [ do-construct-name ]
CONTINUE
```

and [] identify optional components of the block DO construct. This statement is looked at in much greater depth in the chapter on control structures.

10.5 Problems

1. Of what data type should the array Seats be in the theatre example?

2. Write a program to read the following data into a 3*3 array.

```
   1     2     3
   4     5     6
   7     8     9
```

Produce totals for each row and column. After successfully doing this modify the program to produce averages for each row and column as well as the totals.

3. Complete the program which calculates a table of Fahrenheit and corresponding Celsius temperatures. Employ the PARAMETER statement for the constant term.

4. Write a program to print out the 12 times table. Typical output would be of the form:

```
   1    *    12    =    12
   2    *    12    =    24
   3    *    12    =    36
```
etc.

5. Write a program that produces a conversion table from litres to pints and vice versa. One litre is approximately 1 3/4 pints. The output should comprise three columns, the middle column comprising integer values from 1 to 10 and the columns on the left and right corresponding the equivalent in pints and litres respectively. This enables the middle column to be treated as either litres or pints and the corresponding equivalent read from the left or right as appropriate.

11
Arrays 3
Further Examples

A good notation has a subtlety and suggestiveness which at times make it seem almost like a live teacher.

Bertrand Russell.

Aims

The aims of the chapter are:–

- To look more formally at the terminology required to precisely describe arrays;
- To introduce ways in which we can manipulate whole arrays and parts of arrays (sections);
- ALLOCATABLE arrays - ways in which the size of an array can be deferred until execution time
- To introduce the concept of array element ordering and physical and virtual memory;
- To introduce ways in which we can initialise arrays using array constructors
- To introduce the WHERE statement and array masking;

11 Arrays 3: Further Examples

11.1 Terminology

Fortran 90 introduces an abundance of array handling features. In order to make the description of these features more precise a number of additional terms have been introduced.

11.1.1 Rank

The number of dimensions of an array is called its rank. A one-dimensional has rank 1, a two-dimensional array has rank 2 and so on.

11.1.2 Bounds

The bounds of an array are the upper and lower limits of the index in each dimension.

11.1.3 Extent

The number of elements along a dimension of an array is called the extent.

```
INTEGER, DIMENSION(-10:15):: Current
```

has bounds -10 and 15 and an extent of 26.

11.1.4 Size

The total number of elements in an array is its size.

11.1.5 Shape

The shape of an array is determined by its rank and its extents in each dimension.

11.1.6 Conformable

Two arrays are said to be conformable if they have the same shape, that is they have the same rank and the same extents in each dimension.

11.2 Whole array manipulation

The examples of arrays so far have shown operations on arrays via array elements. One of the significant improvements in Fortran 90 is the ability to manipulate arrays as whole objects. This allows arrays to be referenced not just as single elements but also as groups of elements. Along with this ability comes a whole host of intrinsic procedures for array processing. These intrinsic procedures are mentioned in Chapter 14, and listed in alphabetical order, with examples in Appendix C.

11.2.1 Assignment

An array name without any indices can appear on both sides of assignment statements and input and output statements. For example a value can be assigned to all the elements of an array in one statement:–

```
REAL, DIMENSION(1:12):: Rainfall
Rainfall=0.0
```

The elements of one array can be assigned to another

```
INTEGER, DIMENSION(1:50)  :: A,B
...
A=B
```

Arrays A and B must be conformable in order to do this.

The following example is **illegal** since X has shape 20 and Z has shape 41.

```
REAL, DIMENSION(1:20)  :: X
REAL, DIMENSION(1:41)  :: Z
X=50.0
Z=X
```

But the following is legal because both arrays are now conformable, i.e. they are both of rank 1 and an extent of 41.

```
REAL , DIMENSION (-20:20)  :: X
REAL , DIMENSION (1:41)    :: Y
X=50.0
Y=X
```

11.2.2 Expressions

All the arithmetic operators available to scalars are available to arrays, but care must be taken because mathematically they may not make sense:–

```
REAL, DIMENSION (1:50)  :: A,B,C,D,E
C=A+B
```

Adds each element of A to the corresponding element of B and assigns the result to C.

```
E=C*D
```

Arrays 3: Further Examples

Multiplies each element of C by the corresponding element of D. This is **not** vector multiplication. To perform a vector dot product there is an intrinsic procedure DOT_PRODUCT, and an example of this is given under the later section on array constructors.

For higher dimensions:–

```
REAL ,DIMENSION (1:10,1:10) :: F,G,H
F=F**0.5
```

Takes the square root of every element of F.

```
H=F+G
```

Adds each element of F to the corresponding element of G.

```
H=F*G
```

Multiplies each element of F by the corresponding element of G The last statement is **not** matrix multiplication. An intrinsic procedure MATMUL performs matrix multiplication, further details are in Appendix C.

The following program demonstrates the power of these whole array features.

```
PROGRAM Temperatures
! This program reads in a grid of temperatures
! (degrees Fahrenheit)at 25 grid references
! and converts them to degrees Celsius
IMPLICIT NONE
REAL, DIMENSION (1:5,1:5) :: Fahrenheit, Celsius
INTEGER :: Longitude, Latitude
!
! Read in the temperatures
!
   DO Latitude=1,5
      PRINT *, 'For Latitude= ',Latitude
      DO Longitude=1,5
         PRINT *, 'Input temperature for Longitude', Longitude
         READ *,Fahrenheit( Longitude, Latitude)
      END DO
   END DO
!
! Conversion applied to all values
   Celsius = 5.0/9.0 * (Fahrenheit - 32.0)
.
END PROGRAM Temperatures
```

11.3 Array Sections

Often it is necessary to access part of an array not the whole, and this is possible with Fortran 90's powerful array manipulation features.

11.3.1 Example 1: Ages

```
INTEGER, DIMENSION (1:20) :: Ages
  Ages(1:5) = 0
! Sets elements 1 to 5 to value 0
  Ages(6:20) = 1
! Sets elements 6 to 20 to value 1
```

11.3.2 Example 2: Examination Results

```
REAL, DIMENSION (1:20,1:10) :: Exam_Results
!
! Rows represent pupils
! Columns represent subjects
!
! Scale column 3 by 2.5 to change original mark out of 40
! to a mark out of 100
!
   Exam_Results(1:20,3) = 2.5 * Exam_Results(1:20,3)
```

11.4 Allocatable Arrays

The sample program in Chapter 9 to read in, sum and average people's weights is restricted to ten people. This means that if we wish to run the program with a different number of people we would have to change the program, albeit only altering the parameter Number_Of_People. This is a common problem in programming and so Fortran 90 provides a facility to handle it. Instead of using:–

```
INTEGER , PARAMETER :: Number_of_People=10
REAL, DIMENSION (Number_of_People) :: Weight
```

The compiler can be told to defer allocating space to this array until execution time, using:–

```
REAL, DIMENSION (:), ALLOCATABLE :: Weight
READ *, Number_of_people
ALLOCATE(Weight(1:Number_of_people))
```

This is done by giving an array a deferred shape by using the ALLOCATABLE attribute and declaring only the rank of the array on the DIMENSION statement. The array can then be allocated space at run time using the ALLOCATE statement.

Arrays 3: Further Examples

Arrays with the allocatable attribute are called deferred shape arrays.

11.4.1 Example 3: Height Above Sea Level

Taking the example in Chapter 10 of heights above sea level measured in feet amend the program so that it converts to metres and then prints out the newly converted heights.

```
PROGRAM Locate
! Variables used
! Conversion - a constant holding the feet to metres
! conversion factor
! Height - used to hold heights above sea level
! Latitude - used to represent latitude
! Longitude - used to represent longitude
! both restricted to integer values
! Number_of_zones - the number of degrees latitude and longitude
IMPLICIT NONE
INTEGER:: Latitude, Longitude, Number_of_Zones
REAL, PARAMETER:: Conversion=0.305
REAL, DIMENSION(:,:), ALLOCATABLE :: Heights
!
! Read in the number of zones and allocate array Heights
! PRINT *, 'Input the number of zones'
  READ *, Number_of_zones
  ALLOCATE (Heights ( 1:Number_of_zones, 1:Number_of_zones) )
!
! Read in heights by longitude
!
  DO Longitude = 1, Number_of_zones
     PRINT *, 'Input heights for longitude = ', Longitude
     READ *, Heights(Longitude, 1:Number_of_zones)
  END DO
!
! Convert heights from feet to metres
!
  Heights = Conversion * Heights
!
! Print out newly converted heights
!
  DO Longitude = 1, Number_of_zones
     PRINT *, Heights(Longitude, 1:Number_of_zones)
  END DO
END PROGRAM Locate
```

11.5 Array Element Ordering

When whole array operations take place there is no stated order in which the evaluations are performed, although Fortran 90 does specify an ordering of the elements of an array. For a lot of operations this is not important, but it is required for the input and output of whole arrays, and for certain intrinsic functions.

Array element ordering states that the elements of an array, regardless of rank, form a linear sequence. The sequence is such that the subscripts along the first dimension vary most rapidly, and those along the last dimension vary most slowly. This is best illustrated by considering for example, a rank two array A defined by:–

```
REAL , DIMENSION(1:4,1:2) :: A
```

A has 8 real elements whose array element order is:–

A(1,1), A(2,1), A(3,1), A(4,1), A(1,2), A(2,2), A(3,2), A(4,2)

i.e. mathematically by column, and not row. This is illustrated graphically below:–

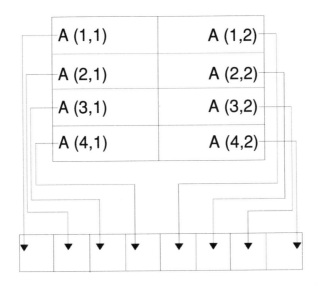

Figure 1

11.5.1 Array Element Ordering and Physical and Virtual Memory

Fortran compilers will store arrays in memory according to the array element ordering scheme above. Whilst the standard says nothing about how this is implemented this generally means in contiguous memory locations.

There will be a limit to the amount of physical memory available on any computer system. To enable problems that require more than the amount of physical memory available to be solved most implementations will provide access to virtual memory, which in reality means access to a portion of a physical disk.

Access to virtual memory is commonly provided by a paging mechanism of some description. Paging is a technique whereby fixed sized blocks of data are swapped between real memory and disk as required.

In order to minimise paging (and hence reduce execution time) array operations should be performed according to the array element order.

Some common page sizes are:–

- DEC VAX – 512 bytes
- DEC Alpha – 8192 bytes
- Intel 80xxx – 4096 bytes

11.6 Array Constructors

Values can be assigned to rank one arrays using an array constructor. For example an integer array with element values (1,3,5,7,9) could be constructed as follows:–

```
INTEGER, DIMENSION (1:5) :: Odd
Odd=(/1,3,5,7,9/)
```

The general form of the array constructor is (/ a list of expressions/) where each expression is of the same type.

To construct arrays of higher rank than one the intrinsic function RESHAPE must be used. (An introduction to Intrinsic functions is given in Chapter 14, and an alphabetic list with a full explanation of each function, is given in Appendix C). To use it in its simplest form:–

Matrix = RESHAPE (Source, Shape)

where Source is a rank one array containing the values of the elements required in the new array, Matrix, and Shape is a rank one array containing the shape of the new array Matrix.

Let's consider the rank one array B=(1,3,5,7,9,11) and we wish to store these values in a rank two array A, such that A is the matrix:–

$$A = \begin{pmatrix} 1 & 7 \\ 3 & 9 \\ 5 & 11 \end{pmatrix}$$

The following code extract is needed:-

```
INTEGER, DIMENSION(1:6)    ::  B
INTEGER, DIMENSION(1:3, 1:2)  ::  A
B = (/1,3,5,7,9,11/)
A = RESHAPE(B,(/3,2/))
```

Note that the elements of the source array B must be stored in the array element order of the required array A.

The following example uses an array constructor and the intrinsic procedure DOT_PRODUCT:–

```
INTEGER , DIMENSION(1:3)  ::  X,Y
INTEGER :: Result
X=(/1,3,5/)
Y=(/2,4,6/)
Result=DOT_PRODUCT(X,Y)
```

and Result has the value 44, i.e. is obtained by the normal mathematical dot product operation, 1*2 + 3*4 + 5*6.

11.7 Masked Array Assignment and the WHERE Statement.

Fortran 90 has array assignment both on an element by element basis and on a whole array basis. There is an additional form of assignment based on the concept of a logical mask.

Consider the earlier example of time zones in chapter 10. The Time array will have values that are both negative and positive. We can then associate with the positive values the concept of east of the Greenwich meridian, and with the negative values the concept of west of the Greenwich meridian, e.g.:–

```
REAL , DIMENSION(-180:180) :: Time
CHARACTER (LEN=1) , DIMENSION(-180:180) :: Direction
   ...
   WHERE (Time > 0.0)
      Direction='E'
   ELSEWHERE (Time < 0.0)
      Direction='W'
   ENDWHERE
   ...
```

Arrays 3: Further Examples

11.7.1 Notes

- The arrays must be conformable, i.e. in our example Time and Direction are the same shape;
- The selective assignment is achieved through the WHERE statement;
- The first array assignment is executed where Time is positive;
- The second array assignment is executed where Time is negative;
- Both the WHERE and ELSEWHERE blocks can be executed therefore;

For further coverage of logical expressions see chapters 15 and 18, and for a formal definition of the WHERE statement see chapter 15.

11.8 Summary

We can now perform operations on whole arrays and part arrays (array sections) without having to refer to individual elements. This shortens program development time, and greatly clarifies the meaning of programs.

Array constructors can be used to assign values to rank one arrays within a program unit. The RESHAPE functions allows us to assign values to a two or higher rank array when used in conjunction with an array constructor.

We have introduced the concept of a deferred shape array. Arrays do not need to have their shape specified at compile time, only their rank. Their actual shape is deferred until run-time. We achieve this by the combined use of the ALLOCATABLE attribute on the variable declaration, and the ALLOCATE statement. This makes Fortran a very flexible language for array manipulation.

11.9 Problems

1. Give the rank, bounds, extents, size of the following arrays:–

```
REAL    , DIMENSION(1:15)       :: A
INTEGER , DIMENSION(1:3,0:4)    :: B
REAL    , DIMENSION(-2:2,0:1,1:4) :: C
INTEGER , DIMENSION(0:2,1:5)    :: D
```

Which two of these arrays are conformable?

2. Write a program to read in five rank one arrays, A, B, C, D, E and then store them as five columns in a rank two array TABLE.

3. Take the first part of problem 2 in chapter 10 and rewrite it using the SUM intrinsic function.

12
Output of Results

Why, sometimes I've believed as many as six impossible things before breakfast.
Lewis Carroll, *Alice through the Looking-Glass*.

All the persons in this book are real and none is fictitious even in part.
Flann O'Brien, *The Hard Life*.

Aims

The aims here are to introduce the facilities for producing neat output, and to show how to write results to a file, rather than to the terminal. In particular

- the A, I, E, F, and X layout or edit descriptors
- the OPEN, WRITE, and CLOSE statements

12 Output

When you have used PRINT * a few times it becomes apparent that it is not always as useful as it might be. The data is written out in a way which makes some sense, but may not be especially easy to read. Real numbers are written out with all their significant places, which is very often rather too many, and it is often difficult to line up the columns for data which is notionally tabular. It is possible to be much more precise in describing the way in which information is presented by the program. To do this, we use FORMAT statements. Through the use of the FORMAT, we can specify how many columns a number should take up, and, where appropriate, where a decimal point should lie. The FORMAT statement has a label associated with it; through this label, the PRINT statement associates the data to be written with the form in which to write it. For example:–

```
PRINT 100, I, XVALUE(I), YVALUE(I)
100 FORMAT(1X,I3,2X,F7.4,2X,F6.4)
```

The label 100 which follows the PRINT statement takes the place of the asterisk we have been using up to now, and links the PRINT with the FORMAT statement. Although the FORMAT follows the PRINT in this example, this is not obligatory. However it is recommended that these statements are always kept together because when errors occur (which they inevitably do!) it will save you time and effort locating the possible source of the error.

The next thing to consider is the FORMAT statement itself, and its contents, i.e. the bits in brackets. The 'X' is used for generating spaces in the output, the 'I' is used when you want to print out an integer, and the 'F' is used when you want to print out real numbers.

12.1 Integers, I format

Integer format is reasonably straightforward, and offers clues for formats used in describing other numbers. I3 is an integer taking three columns. The number is right justified, a bit of jargon meaning that it is written as far to the right as it will go, so that there are no trailing or following blanks

```
 423
 -22
   9
```

are all right justified integers (in I3). Note that the minus sign counts as part of the number, and takes up one of the three columns we have specified here. The only problem with right justification is that, when you try to write something which looks useful, such as

```
THERE ARE 3 IMAGES AVAILABLE FOR PROCESSING
```

the integer part must be assigned a fixed size. In the above example, we might have given the field I2 to the number of images. This is fine when there are 0 to 9 images, but when there are more than 9, the message might read:–

```
THERE ARE12 IMAGES AVAILABLE FOR PROCESSING
```

which does not look very tidy, and is more difficult to read. Another alternative is to give that a very large field, say I10. But this would look extremely disjointed:–

```
THERE ARE           3 IMAGES AVAILABLE FOR PROCESSING
```

The definition of an integer format is therefore the letter I, followed by a positive integer number, e.g.

```
I1
I10
I3
I6
I12
```

12.2 Reals, F format

The F format can be seen as an extension of the integer format. But here we have to deal with the decimal point. The form of the F format specifies where the decimal point will occur, and how many digits follow it. Thus, F7.4 means that there are 4 digits after the point, in a total field width of 7 digits. Since the decimal point is also written out, there may be up to 2 digits before the decimal point. As in the case of the integer, any minus sign is part of the number, and would take up one column. Thus, the format F7.4 may be used for numbers in the range

$$-9.9999 \quad \text{to} \quad 99.9999$$

Let us look at the last example more closely. When a number is written out, it is rounded; that is to say, if we write out 99.99999 in an F7.4 format, the program will try to write out 100.0000! This is bad news, since we have not left enough room for all those digits before the decimal point. What happens? Asterisks will be printed. In the example above, a number out of range of the format's capabilities would be printed as:–

```
*******
```

What would a format of F7.0 do? Again, seven columns have been set aside to accommodate the number and its decimal point, but this time no digits follow the point:–

```
    99.
-21375.
```

are examples of numbers written in this format. With an F format, there is no way of getting rid of the decimal point.

The numbers making up the parts of the descriptors must all be positive integers. The definition of a real format is therefore; F followed by two integer numbers, separated by a decimal point. The first integer must exceed the second, and the second must be greater than or equal to zero. The following are valid examples:–

    F4.0
    F6.2
    F12.2
    F16.8

but these are *not* valid:–

    F4.4
    F6.8
    F-3.0
    F6
    F.2

12.3 Reals, E format

The exponential or scientific notation is useful in cases where we need to provide a format which may encompass a wide range of values. If likely results lie in a very wide range, we can ensure that the most significant part is given. It is possible to give a very large F format, but alternatively, the E format may be used. This takes a form such as

    E10.4

which looks something like the F, and may be interpreted in a similar way. The 10 gives the total *width* of the number to be printed out, that is the number of columns it will take. The number after the decimal point indicates the number of positions to be written after the decimal point; since all exponent format numbers are written so that the number is between 0.1 and 0.9999..., with the exponent taking care of scale shifts, this implies that the first four significant digits are to be printed out.

Taking a concrete example, 1000 may be written as 10**3, or as 0.1 * 10**4. This gives us the two parts; 0.1 gives the significant digits (in this case only one significant digit), while the 10**4 gives the exponent, namely 4 or +4. In a form that looks more like Fortran, this would be written .1E+04 where the E+04 means 10**4.

There is a minimum *size* for an exponential format. Because of all the extra bits and pieces it requires (the decimal point, the sign of the entire number, the sign of the exponent, the magnitude of the exponent and the E), the width of the number, less the number of significant places should not be less than 6. In the example given above, E10.4 meets this requirement. When the exponent is in the range zero to 99, the E will be printed as part of the number; when the exponent is greater, the E is dropped, and its place is taken by a larger value; however, the sign of the exponent is always given, whether it is positive or negative. The sign of the whole number will usually only be given when it is negative. This means that, if the numbers are always positive, the *rule of six* given above can be modified to a *rule of five*. It is safer to allow six places over, since, if the format is insufficient, all you will get are asterisks.

The most common mistake with an E format is to make the edit descriptor too small, so that there is insufficient room for all the *padding* to be printed. Formats like E8.4 just don't work (on output anyway). The following are valid E formats on output:–

```
E9.3
E11.2
E18.7
E10.4
```

but the following would not be acceptable as output formats, for a variety of reasons

```
E11.7
E6.3
E4.0
E10
E7.3
```

The first example in this chapter could be rewritten to use E format as:–

```
     PRINT 100,I,XVALUE(I),YVALUE(I)
100  FORMAT(1X,I3,E10.4,3X,E10.4)
```

This would be of use when we are unsure about the exact range of the numbers to be printed.

12.4 Spaces

There is a shorthand way of expressing spaces (or blanks) in your output — the X descriptor, e.g.

```
     PRINT 100, ALPHA,BETA
100  FORMAT(1X,F10.4,10X,F10.3)
```

The 10X is read rather like any of the other format elements — logically it should have been X10, to correspond to I10 or F10.4, but that would be allowing intuition to run away with you. Clearly the X3J3 committee felt it important that Fortran should have inconsistencies, just like a natural language.

There are other ways of achieving the same thing — having a large space delimited by apostrophes, or by manipulating character variables. What you use is your choice. Remember that these blanks are in addition to any generated as a result of the leading blanks on numbers (if any are present). If you wish to leave a single space, you must still precede the X by a number (in this case, 1); simply writing X is illegal. The general form is therefore; positive integer followed by X.

12.5 Alphanumeric or character format, A

This is perhaps the simplest output of all. Since you will already have declared the length of a character variable in your declarations:-

    CHARACTER (10) :: B

when you come to write out B, the length is known — thus you need only specify that a character string is to be output:-

    PRINT 100,B
    100 FORMAT(A)

If you feel you need a little extra control, you can append an integer value to the A, like A10 (A9 or A1) and so on. If you do this, only the first 10 (9 or 1) characters are written out; the remainder are ignored. Do note that 10A1 and A10 are not the same thing. 10A1 would be used to print out the first character of ten character variables, while A10 would write out the first ten characters of a single character variable. The general form is therefore just A, but if more control is required, this may be followed by a positive integer.

Within a FORMAT statement you may also write out anything within apostrophes. These strings of characters will be written out with no modification; e.g.

    PRINT 100,A
    100 FORMAT(' THE ANSWER IS',F10.4)

will be written out as something like:-

    THE ANSWER IS   12.3457

A partial program segment to output a 3 column table with an informative heading could be:-

    PRINT *,' NUMBER : X READING : Y READING'
    DO I=1,100
       PRINT 100,I,XVALUE(I),YVALUE(I)
       100   FORMAT(1X,I3,3X,E10.4,3X,E10.4)
    END DO

12.6 Common mistakes

It must be stressed that an integer can only be printed out with an I format, and a real with an F (or E) format. You cannot mix integer and F, or real and I. If you do, unpredictable results will follow. There are (at least) two other sorts of errors you might make on writing out a value. You may try to write out something which has never actually been assigned a value; this is termed an indefinite value. You may find that the letter I is written out. In passing, note that many loaders and link editors will preset all values to zero — i.e. unset (indefinite) values are actually set to zero. On better systems there is generally some way of turning this facility off, so that undefined is really indefinite. More often than not, indefinite values are the result of mis-typing, rather than never setting values. It is not uncommon to type O for 0, or 1 for either I or l. The other likely error is to try to print out a value greater than that which the machine can calculate — *out of range* values. Some machines will print out such values as R; some machines will actually print out something which looks right, and such *overflow* and *underflow* conditions can go unnoticed. Be wary.

12.7 OPEN (and CLOSE)

One of the particularly powerful features of Fortran is the way it allows you to manipulate files. Up to now, most of the discussion has centred on reading from and writing to the terminal. It is possible to read and write to one or more files. This is achieved using the OPEN, WRITE, READ and CLOSE statements. We will consider *reading* from files in a later chapter, and concentrate on *writing* in this chapter.

12.7.1 The OPEN statement

This statement sets up a file for either reading or writing. A typical form is:–

```
OPEN (UNIT=1,FILE='DATA')
```

The file will be known to the operating system as DATA (or will have DATA as the first part of its name), and can be written to by using the UNIT number. This statement should come *before* you first read from or write to the file DATA.

It is not possible to write to the file DATA directly; it must be referenced through its unit number. Within the Fortran program you write to this file using a statement like

```
WRITE(UNIT=1,FMT=100) XVAL,YVAL
```

or

```
WRITE(1,100) XVAL,YVAL
```

These two statements are equivalent. Besides opening a file, we really ought to CLOSE it when we have finished writing to it:

```
CLOSE(UNIT=1)
```

In fact, on many systems it is not obligatory to OPEN and CLOSE all your files. Almost certainly, the terminal will not require this, since INPUT and OUTPUT units will be there by default. At the end of the job, the system will CLOSE all your files. Nevertheless, explicit OPEN and CLOSE cannot hurt, and the added clarity generally assists in understanding the program.

The following program segment contains all of the above statements.

```
PROGRAM Redo
...
   OPEN (UNIT=1,FILE='DATA')
   ...
   DO I=1,100
      READ (UNIT=1,FMT=200) X(I)
      200 FORMAT(E10.3)
      SUM = SUM + X(I)
   END DO
   ...
   CLOSE(1)
   ...
END PROGRAM Redo
```

12.7.2 Writing

PRINT is always directed to the file OUTPUT; in the case of interactive working, this is the terminal. This is not a very flexible arrangement. WRITE allows us to direct output to any file, including *OUTPUT*. The basic form of the WRITE is

```
WRITE(6,100) X,Y,Z
```

or

```
WRITE(UNIT=6,FMT=100) X,Y,Z
```

The latter form is more explicit, but the former is probably the one most widely used. We have an example here of the use of positionally dependent parameters in the first case and equated keywords in the second. With the exceptions of the PRINT statement and the READ * form of the READ, all of the input/output statements allow the unit number and the format labels to be specified either by an equated keyword (or specifier), or in a positionally dependent form. If you use the explicit UNIT= and FMT= it does not matter what order the elements are placed in, but if you omit these keywords, the unit number must come first, followed by the format label.

UNIT=6 means that the output will be written to the file given the unit number 6. In the next chapter we will cover the way in which you may associate file names and unit numbers, but, for the moment, we will assume that the default is being used. The name of the file, as defined by the system, will depend on the particular system you use; a likely name is something like DATA06, TAPE6, or FILE0006. One *easy* way to find out (apart from asking someone), is to create such a file from a program, and then look at the names of your files after the program has finished. A great many of computing's minor complexities can be clarified by simple experimentation.

FMT=100 simply gives the label of the format to be used.

The overworked asterisk may be used, either for the unit, or for the format:–

UNIT=* will write to OUTPUT (the terminal), and

FMT=* will produce output controlled by the list of variables, often called *list directed output*.
The following three statements are therefore equivalent:–

```
WRITE(UNIT=*,FMT=*) X,Y,Z
WRITE(*,*) X,Y,Z
PRINT*,X,Y,Z
```

There are other controls possible on the WRITE, which will be elaborated later.

12.8 Repetition

Often we need to print more than one number on a line and want to use the same layout descriptor. Consider the following:–

```
PRINT 100,A,B,C,D
```

If each number can be written with the same layout descriptor, we can abbreviate the FORMAT statement to take account of the pattern:–

```
100 FORMAT(1X,4F8.2)
```

is equivalent to:–

```
100 FORMAT(1X,F8.2,F8.2,F8.2,F8.2)
```

as you might anticipate. If the pattern is more complex, we can extend this approach:–

```
PRINT 100,I,A,J,B,K,C
100 FORMAT(1X,3(I3,F8.2))
```

Bracketing the description ensures that we repeat the whole entity:–

```
100 FORMAT(1X,3(I3,F8.2))
```

is equivalent to:–

```
100 FORMAT(1X,I3,F8.2, I3,F8.2, I3,F8.2)
```

Repetition with brackets can be rather more complex. In order to give some overview of formatted Fortran output, it is helpful to delve a little into the history of the language. Many of the attributes of Fortran can be traced back to the days of single user mainframes (with often a fraction of the power of many contemporary micro-computers and work-stations). These would generally take input from punched cards (the traditional 80-column Hollerith card), and would generate output on a line printer. In this sort of environment, the individual punched card had a significance which lines in a

Output

file do not have today. Each card could be seen as a single entity — a physical record unit. The *record* was seen as an element of subdivision within a file. Even then, there was some confusion between the notion of physical records and files split into logically distinct sub-units, since these sub-units might also be termed records. The present Fortran standard merely says that a record *does not necessarily correspond to a physical entity*, although *a punched card is usually considered to be a record*. This leaves us sitting at our terminals in a bemused state, especially since we may have no idea what a punched card looks like (an ideal state of affairs!)

It is important to have some notion of a record, since most of the formal definitions dealing with output (and input) are couched in terms of records. Every time an input or output statement is executed your nominal position in the file changes. If we think in terms of individual records (which may be cards), the notions of *current, preceding* and *next* record seem fairly straightforward. The current record is simply the one we have just read or written, and the other definitions follow naturally.

The situation becomes less clear when we realise that a single output statement may generate many lines of output.

```
    WRITE(UNIT=6,FMT=101)    A,B,C
101 FORMAT(1X,F10.4)
```

writes out three separate lines. Looking at the output alone, there is no way to distinguish this from the output generated by:-

```
    WRITE(UNIT=6,FMT=101)    A
    WRITE(UNIT=6,FMT=101)    B
    WRITE(UNIT=6,FMT=101)    C
101 FORMAT(1X,F10.4)
```

In the latter case we would probably be happy to consider each line a *record*, although in the previous example we might swither between considering all three lines (generated by a single statement) a single record or three records. Consider the first of these two examples more closely; each time the format is exhausted — that is to say, each time we run out of format description, we start again on a new line (a new record). A new record is begun as each F10.4 is begun. The correct interpretation is therefore that three records have been written.

The same sort of thing happens in more complex FORMAT statements:-

```
    WRITE(UNIT=6,FMT=105)    X,I,Y
105 FORMAT(1X,F8.4,I3,(F8.4))
```

would write out a single record containing a real, an integer and a real. Using the same format statement with WRITE (UNIT=6, FMT=105) X,I,Y,Z would write out two records. The first containing the values of X, I and Y, the second containing only Z. If there were still more values

```
WRITE(UNIT=6,FMT=105) X,I,Y,Z,A
```

would print out three records. The group in brackets — the (F8.4) — is repeated until we run out of items.

12.9 Some more examples

Since it is the last open bracket which determines the position at which the format is repeated, simply writing:–

```
WRITE(UNIT=6,FMT=100)   A,I,B,C,J
100 FORMAT(1X,F8.4,I3,F8.2)
```

would imply that A, I and B would be written on one line, then, returning to the last open brackets, (in this case the only open brackets), a new record (or line) is begun to write out C and J. A statement like:–

```
100 FORMAT(1X,(F8.4),I3,F8.2)
```

would return to the (F8.4) group, and then continue to the I3 and F8.2 before repeating again (if necessary). The same thing happens if the (F8.4) had no brackets around it. On the other hand:–

```
100 FORMAT(1X,(F8.4),I3,(F8.2))
```

contains superfluous brackets around the F8.4, since the repeat statement will never return to that group. Are you confused yet? This seems all very esoteric, and really, we have only hinted at the complexity which is possible. It is seldom that you have to create complex FORMAT statements, and clarity is far more important than brevity.

When patterned or repeated output is used, we may want to stop when there are no more numbers to write out. Take the following example:–

```
WRITE(UNIT=1,FMT=100) A,B,C,D
100 FORMAT(1X,4(F6.1,','))
```

This will give output which looks like:–

```
  37.4,   29.4,   14.2,   -9.1,
```

The last comma should not be there. We can suppress these unwanted elements by using the colon:–

```
100 FORMAT(1X,4(F6.1:','))
```

which would then give us:–

```
  37.4,   29.4,   14.2,   -9.1
```

Output

Since we run out of data at the fourth item, D, the output following is not written out. It is a small point, but it does look a lot tidier. There are other ways of achieving the same thing.

This helps to illustrate another point, namely that you may have formats which are more extensive than the lists which reference them:–

```
WRITE(UNIT=1,FMT=100)  A,B,C
WRITE(UNIT=1,FMT=100)  X,Y
100 FORMAT(1X,6F8.2)
```

Both WRITE statements use the format provided, although they write out different amounts of data, and neither uses up the whole format.

12.10 Implied DO loops

In reading and writing it is possible to use more compact ways of indicating that an array is being referenced, since it is often rather tedious to declare each element which is involved, e.g.

```
WRITE(UNIT=6,FMT=100)Y(1),Y(2),Y(3),Y(4)
100 FORMAT(1X,F8.4)
```

Clearly we could improve this slightly by making it into a loop:–

```
DO I=1,4
   WRITE(UNIT=6,FMT=100)  Y(I)
   100 FORMAT(1X,F8.4)
END DO
```

and equally we can simplify this to:–

```
WRITE(UNIT=6,FMT=100) (Y(I),I=1,4)
100 FORMAT(1X,F8.4)
```

where the DO loop is subsumed into the expression (the syntax is just the same as for the counter part of a DO loop, with the same rules for starting, ending and incrementing). An alternative, and yet more compact form is:–

```
REAL , DIMENSION(4) ::  Y
   ...
   ...
   WRITE(UNIT=6,FMT=100)Y
   100 FORMAT(1X,F8.4)
```

In all these cases, the output would be the same — four numbers printed out on separate lines. The FORMAT statement is controlling this layout. Changing the format to:–

```
100 FORMAT(1X,4F8.4)
```

would change the layout in three out of the four cases outlined. Which three? Even two (or more) dimensional arrays can be written out or read in by implied loops.

```
REAL , DIMENSION(10,10) :: Y
   NROWS=6
   NCOLS=7
   DO J=1,NROWS
      WRITE(UNIT=6,FMT=100)(Y(I,J),I=1,NCOLS)
      100 FORMAT(1X,10F10.4)
   END DO
```

may be written:–

```
WRITE(UNIT=6,FMT=100)((Y(I,J),I=1,NCOLS),J=1,NROWS)
```

or even as:–

```
WRITE(UNIT=6,FMT=100) Y
```

There are two points to note with this last example. Firstly, this is a whole array reference, and so the entire contents of the array will be written; there is no scope for fine control. Secondly, the order in which the array elements are written is according to Fortran's array element ordering, i.e. the first subscript varying 1 to 10 (the array bound), with the second subscript as 1, then 1 to 10 with the second subscript as 2 and so on; the sequence is

```
Y(1,1)     Y(2,1)     Y(3,1)        Y(10,1)
Y(1,2)     Y(2,2)     Y(3,2)        Y(10,2)
   .
   .
   .
Y(1,10)    Y(2,10)                  Y(10,10)
```

Another feature to note is that we can generate values from within a WRITE statement:–

```
WRITE(UNIT=6,FMT=101)(I,I=0,9)
101 FORMAT(1X,10I3)
```

would produce a line like:–

```
  0  1  2  3  4  5  6  7  8  9
```

12.11 Formatting for a line-printer

There is one extension to format specifications which is relevant to line-printers. Fortran defines four special characters which have a particular effect on standard line-printers. They have an effect when they occur in the first character position of a line. This means that a line-printer which is not under your immediate control can be used to produce neat output, by sending a file to be printed on it. This has a variety of names including *spooling*, *queueing* and *routing* depending on the system. You should check with your local system for the exact mechanism to achieve this.

The special characters are +, 0, 1 and blank. To be used, they must be the first character of the output in each line — as if they were to be printed in column 1. In fact, a standard line printer never prints a character that occurs in column 1 at all.

Whenever a WRITE statement is begun, the printer *advances* to a new record; i.e. a new line is begun before any data is transferred. If the first character is a *special character,* then this will be interpreted by the line-printer. If the first character to be printed is a blank, the printer continues printing on that line. The first character is also known as the *carriage control character*.

The blank is a *do nothing special* control. It signifies that the line is to be printed as it is.

The zero indicates that you wish to leave an extra line; this is often useful in spacing out results to make the output more readable.

The 1 makes the output skip down to the top of the next page. This is clearly useful for separating logically distinct chunks of output. If you obtain a line printer listing of your compiled program, each segment will start at the top of a new page.

The plus is a *no advance* or *overprint* character. It suppresses the effect of the line advance which a WRITE generates. No new line is begun and the previous line is over-printed with the new. Overprinting can be useful especially when you wish to print out grey scale maps but its use is rather restricted. In particular, it can be a dangerous control character. If you have a format starting with a plus in a loop, you can make the printer overprint again and again and again .... and again and again, until it has hammered itself into a pulp. This is not a good idea.

Similarly, accidental use of the 1 as a control character in a loop will give you lots of blank pages. It is just a bit embarrassing to be presented with a six inch stack of paper which is (almost) blank, because you had a 1 repeatedly in column 1.

12.11.1 Mechanics of carriage control

The following are all quite reasonable ways of generating the blank in column 1:-

```
WRITE(UNIT=6,FMT=100)A
100 FORMAT(' ',F10.4)
```

or

```
WRITE(UNIT=6,FMT=100)A
100 FORMAT(1X,F10.4)
```

or

```
WRITE(UNIT=6,FMT=100)A
100 FORMAT(' THE ANSWER IS ',F10.4)
```

Note, however, that

```
WRITE(UNIT=6,FMT=100)A
100 FORMAT(F8.4)
```

could result in problems. If A contained the value 100.2934, the result on a line printer would be

```
00.2934
```

printed at the top of a new page. The 1 is taken as carriage control, and the rest of the line then printed.

Accidentally printing zeros in column 1 is a little more difficult, but:–

```
WRITE(UNIT=6,FMT=100)I
100 FORMAT(I1)
```

might just do it. Don't.

Remember that this only applies to line printer output, and not to the terminal. Since Fortran only defines 4 characters as carriage control, you will find that anything else in column 1 will give unpredictable results. On some systems, a fair number of alternatives may be defined by the installation, and they may do something useful. On other systems, they may do something, but they may also fail to print the rest of the line. This can be very perplexing. Beware.

12.11.2 Generating a new line, on both line-printers and terminals.

There are several ways of generating new lines, other than with a 0 in column one of your line printer output. A more general approach, which works on terminals and also line printers, is through the oblique or slash, /. Each time this is encountered in a FORMAT statement, a new line is begun.

```
   PRINT 101,A,B
101 FORMAT(1X,F10.4/1X,F10.4)
```

would give output like:–

```
100.2317
-4.0021
```

This is the same as (F10.4) would have given, but clearly this opens up lots of possibilities for formatting output more tidily:–

```
      PRINT 102,NVAL,XMAX,XMIN
  102 FORMAT(' NUMBER OF VALUES READ IN WAS: ',I10/ &
             ' MAXIMUM VALUE IS: ',F10.4/            &
             ' MINIMUM VALUE IS: ',F10.4)
```

which may be easier to read than using only one line, and is certainly more compact to write than using three separate print statements. It is not necessary to separate / by commas, although if you do nothing catastrophic will happen.

You may also begin a format description with a /, in order to generate an extra line, or even generate lots of lines with lots of slashes; e.g.

```
      WRITE(UNIT=6,FMT=103)A,B
  103 FORMAT(//1X,F10.4,4(/),1X,F10.4)
```

will leave two lines before printing A, and then will generate 4 new lines before writing B (i.e. there will be three lines between A and B — the fourth new line will contain B). While a slash by itself, or with another slash, does not have to be separated by commas from other groups, a more complex grouping, 4(/), does have to have commas and brackets to delimit it.

12.12 Summary

• You have been introduced in this chapter to the use of format or layout descriptors which will allow greater control over output.

• The main features are the I format for integer variables, the E and F formats for real numbers, and the A format for characters. In addition the X, which allows insertion of spaces, has been introduced. • Output can be directed to files as well as to the terminal, through the WRITE statement.

• The WRITE, together with the OPEN and CLOSE statements, also introduces the class of Fortran statements which use equated keywords, as well as positionally dependent parameters.

• The FORMAT statement and its associated layout or edit descriptor are powerful, and allow repetition of patterns of output (both explicitly and implicitly).

• When output is to be directed to a line-printer, there are four characters defined that allow reasonable control over the layout. Care must to be taken with these characters, since it is possible to decimate forests with little effort.

12.13 Problems

1. Write a program to produce the following kind of conversion table:–

CELSIUS	TEMPERATURE	FAHRENHEIT
-73.3	-100	-148.0

etc.

The centre column of the table should start at –100 and go up to +100. Use F format to print out the values of CELSIUS and FAHRENHEIT, whilst the central column should use I format.

Fahrenheit temperature = (Celsius/5) * 9 +32
Celsius temperature = 5 * (Fahrenheit–32)/9

2. Write a litres and pints conversion program to produce a similar kind of output to the above. Start at 0, and make the central column go up to 50. One pint is 0.568 litres.

3. Information on car fuel consumption is usually given in miles per gallon in Britain and the US, and in litres per 100 Km in Europe. Just to add an extra problem US gallons are 0.8 Imperial gallons.

Prepare a table which allows conversion from either US or Imperial fuel consumption figures to the metric equivalent. Use the PARAMETER statement where appropriate.

1 Imperial gallon = 4.54596 litres
1 mile = 1.60934 kilometres.

4. Modify any of the above to write to a file rather than the terminal. What changes are required to produce a general output which will be suitable for both the terminal and a line printer? Is this degree of generality worthwhile?

5. To demonstrate your familiarity with formats, re-format questions 1, 2 or 3 to use E formats, rather than F (or vice versa).

6. Modify the temperature conversion program to produce output suitable for a line-printer. Use the local operating system commands to send the file to be printed.

7. Repeat for the litres and pints program.

8. What features of Fortran reveal its evolution from punched card input?

9. Try to create a real number greater than the maximum possible on your computer — write it out. Try to repeat this for an integer. You may have to exercise some ingenuity.

10. Check what a number too large for the output format will be printed as on your local system – is it all asterisks?

11. Write a program which stores litres and corresponding pints in arrays. You should now be able to control the output of the table (excluding headings — although this could be done too) in a single WRITE or PRINT statement. If you don't like litres and pints, try some other conversion (sterling to US dollars, leagues to fathoms, Scots miles to Betelgeusian pfnings). The principle remains the same.

13
Reading in data

Winnie-the-Pooh read the two notices very carefully,
first from left to right, and afterwards,
in case he had missed some of it, from right to left.

A A Milne, *Winnie-the-Pooh*.

For Madmen Only

Hermann Hesse, *Steppenwolf*.

Aims

The aims of this chapter are to introduce some of the ideas involved in reading data into a program. In particular, using the following:–

- reading from fixed fields
- integers, reals and characters
- blanks — nulls or zeros?
- READ — extensions
    - error handling on input
- OPEN — associating unit numbers and file names
    - CLOSE
    - REWIND
    - BACKSPACE

13 Reading in Data

13.1 Fixed fields on input

All the formats described earlier are available, and again they are limited to particular types. Integers may only be input by the I format, reals with F and E, and character (alphanumeric) with A.

13.1.1 Integers, the I format

Integers are read in with the I edit descriptor. While, on output, integers appear right justified, on input they may appear anywhere in the field you have delimited. Blanks (by default) are considered not to exist, for the purpose of the value read, although they do contribute to the field width. Apart from the digits 0 to 9, the only other characters which may appear in an integer field are – and +.

```
READ(UNIT=*,FMT=100)  I,J,K
100 FORMAT(3I4)
```

with the following values:–

```
2   -4 0+  21
```

would result in the values 2, – 40 and 21 being assigned to I, J and K respectively.

13.1.2 Reals, the F and E formats

Real numbers may be input using either the E or F format, whether or not the E descriptor is present in the field. Again, we define a width, and, as with output, the number of places after the point:–

```
F10.4         E12.3         F6.0         E10.0
```

However, if the point is already present in the value being input, this overrides the definition in the format. Again, blanks are treated as null values.

```
READ(UNIT=*,FMT=100)  A,B,C
100 FORMAT(F10.6,E12.6,F6.0)
```

with

```
1234567890   14                 4   .
```

results in A taking the value 1234.56789, B taking the value 14.0, and C the value 4.

The absence of the E in the field for B has no adverse affect. As a general rule, it is best to retain the decimal point with real numbers, just as a precaution. Sometimes it is difficult to line up fields properly, and the first sign of trouble may be finding two decimal points in the one field, which will generate an error message, e.g. consider:–

Reading in Data

```
      READ(UNIT=*,FMT=101)  A,B,C,X,Y,Z
  101 FORMAT(6F5.2)
```

If this was in a loop to read the following values:-

```
2.0   3.0   13.0  6.1   0.9   0.2
12.0  62.   9.    -4.2  2.9   -3.7
```

The second time the READ was used, you would get an error (can you see why and where?)

An exponential format number (which may be read in F or E formats) can take a number of different forms. The most obvious is the explicit form:-

```
-1.2E-4
```

where all the components of the value are present — the significant digits to the left of the E, the E itself, and the exponent to the right. We can drop almost any two of these three components, and therefore:-

```
-1.2
-1.2E
-1.2-4
-4
```

are all valid values. Only the first two are interpreted as the same numerical value, and just giving the exponent part would be interpreted by the format as just giving the significant digits. If the exponent is to be given, there must be some significant digits also. It is not even enough to give the E and assume that the program will interpret this as 10 to the power *exponent* :-

```
E-4
```

is not an acceptable exponential format value, although:-

```
1E-4
```

would be.

There are opportunities for confusion with E formats.

```
      READ(UNIT=*,FMT=102)  X,Y
  102 FORMAT(2E10.3)
```

with:-

```
10.23 -2
```

This would be interpreted as X taking the value 10.23E-2 and Y taking the value 0.0, while with

```
102 FORMAT(2F8.3)
```

X would be 10.23, and Y would be –2.0.

Although the decimal point may also be dropped, this may generate confusion too. While:–

```
4E3
45
45E-4
45-4
```

are all valid forms, if an E format is used, a special conversion takes place. A format, like E10.8, when used with integral significant digits (no decimal point), uses the 8 as a *negative* power of 10 scaling, e.g.

```
3267E05
```

converts to

```
3267*10**-8*10**5
```

or

```
3267*10**3
```

or

```
3.267.
```

Therefore, the interpretation of, say, 136, read in E format, would depend on the format used:–

Value	Format	Interpretation
136	E10.0	136.0
136	E10.4	136.0*10**–4 or 0.0136
136	E10.10	136.0*10**–10 or 0.0000000136
136.	any above	136.0

Reading in Data

One implication of all this is that the format you use to input a variable may not be suitable to output that same variable.

13.2 Blanks, nulls and zeros

You can control how Fortran treats blanks in input through two special format instructions, BN and BZ. BN is a shorthand form of *blanks become null,* that is, a blank is treated as if it was not there at all. BZ is therefore *blanks become zeros.*

As we have already seen, 1 4 (i.e. the two digits separated by a blank) read in I3 format would be read as 14; similarly, 14 (one-four-blank) is also 14 when the BN format is in operation. All of the blanks are ignored for the purposes of interpreting the number. They help to create the width of the number, but otherwise contribute nothing. This is the default, which will be in operation unless you specify otherwise.

The BZ descriptor turns blanks into zeros. Thus, 1 4 (one-blank-four) read in I3 format is 104, and 14 (one-four-blank) is 140.

There is one place where we must be very careful with the use of the BZ format — when using exponent format input. Consider:–

```
5.321E+02
```

read in (BZ,E10.3) format. We have specified a field which is ten characters wide, therefore the blank in column 10, which follows the E+02, is read as a zero, making this E+020. This is probably not what was required.

13.3 Characters

When characters are read in, it is sufficient to use the A format, with no explicit mention of the size of the character string, since this size (or length) is determined in the program by the CHARACTER declaration. This implies that any *extra* characters would not be read in. You may however read in less:–

```
CHARACTER (10) :: LIST
   .
   .
   READ(UNIT=5,FMT=100)LIST
   100 FORMAT(A1)
```

would read only the first character of the input. The remaining 9 characters of LIST would be set to blank.

The notion of blanks as nulls or zeros has no meaning for characters. The blank is a legitimate character, and is treated as meaningful, completely distinct from the notion of a null or a zero.

13.4 Skipping spaces and lines

The X format is also useful for input. There may be fields in your data which you do not wish to read. These are easily omitted by the X format:–

```
READ(UNIT=5,FMT=100) A,B
100 FORMAT(F10.4,10X,F8.3)
```

Similarly, you can *jump over* or ignore entire records, by using the oblique. Do note however, that

```
READ(UNIT=5,FMT=100) A,B
100 FORMAT(F10.4/F10.4)
```

would read A from one line (or record) and B from the next. To omit a record between A and B, the format would need to be:–

```
100 FORMAT(F10.4//F10.4)
```

Another way to skip over a record is:–

```
READ(UNIT=5,FMT=100)
100 FORMAT()
```

with no variable name at all.

13.5 Reading

As you have seen already, reading, or the input of information, is accomplished through the READ statement. We have used:–

```
READ *,X,Y
```

for list directed input from the terminal, and:–

```
READ(UNIT=5,FMT=100) X,Y
```

for formatted input also from the terminal. These forms may be expanded to

```
READ(UNIT=*,FMT=*) X,Y
```

or

```
READ(UNIT=*,FMT=100) X,Y
```

for input from the terminal, or to

```
READ(UNIT=5,FMT=*) X,Y
```

Reading in Data

or

```
READ(UNIT=5,FMT=100) X,Y
```

when we wish to associate the READ statement with a particular unit number (or format label, for formatted input). As with the WRITE statement, these last two READ statements may be abbreviated to

```
READ(5,*) X,Y
```

and

```
READ(5,100) X,Y
```

13.6 File manipulation again

The OPEN and CLOSE statements are also relevant to files which are used as input, and they may be used in the same ways. Besides introducing the notion of manipulating lots of files, the OPEN statement allows you to change the default for the treatment of blanks. The default is to treat blanks as null, but the statement BLANK='ZERO' changes the default to treat blanks as zeros. There are other parameters on the OPEN, which are considered elsewhere.

Once you have OPENed a file, you may not issue another OPEN for the same file until it has been CLOSEd, except in the case of the BLANK= parameter. You may change the default back again with:-

```
OPEN(UNIT=10,FILE='Example.dat')
READ(UNIT=10,FMT=100) A,B
...
OPEN(UNIT=10,FILE='Example.dat',BLANK='ZERO')
READ(UNIT=10,FMT=100) A,B
```

This implies that, within the same input file, you may treat some records as blank for null, and some as blank for zero. This sounds very dangerous, and would be better done by manipulating individual formats if it had to be done at all.

Given that you may write a file, you may also *rewind* it, in order to get back to the beginning. The syntax is similar to the other commands:-

```
REWIND(UNIT=1)
```

This often comes in useful as a way of providing backing storage, where intermediate data can be stored on file and then used at a later part of the processing.

The notion of records in Fortran input and output has been introduced. If you are confident in your understanding of this ambiguous and nebulous concept, you can *backspace* through a file, using the statement

```
BACKSPACE(UNIT=1)
```

which moves back over a single record on the designated file. There is no point in trying to BACKSPACE or REWIND input, if that input is the terminal.

13.7 Errors when reading

In discussing some aspects of input, it has been pointed out that errors may be made. Where such errors are noticed, in the sense that something illegal is being attempted, there are two options

- print a diagnostic message, and allow correction of the mistake
- print a diagnostic message, and terminate the program

The only time that the first makes sense is when you are interacting with a program at a terminal. Some Fortran implementations provide correction facilities in a case like this, but most do not.

We will look in chapter 21 at how we handle errors in input data.

13.8 Summary

• Values may be read in from the terminal or from another file through fixed formats.

• Much of the structure of input format statements is very similar to that of the output formats. Broadly speaking, data written out in a particular format may be read in by the same format. However, there is greater flexibility, and quite a variety of forms can be accepted on input.

• A key distinction to make is the interpretation of blanks, as either nulls or zeros; alternative interpretations can radically alter the structure of the input data

• Fortran allows file names to be associated with unit numbers through the OPEN statement. This statement allows control of the interpretation of blank, although this can also be done through the BN and BZ formats.

• Files may also be manipulated through REWIND and BACKSPACE.

13.9 Problems

1. Write a program that will read in two reals and one integer, using

```
FORMAT(F7.3,I4,F4.1)
```

and that, in one instance treats blanks as zeros, and in the second treats blanks as nulls. Use PRINT *, to print the numbers out immediately after reading them in. What do you notice? Can you think of instances where it is necessary to use one rather than the other?

2. Write a program to read in and write out a real number using

FORMAT(F7.2)

What is the largest number that you can read in and write out with this format? What is the largest negative number that you can read in and write out with this format? What is the smallest number, other than zero, that can be read in and written out?

3. Rewrite two of the earlier programs that used READ,* and PRINT,* to use FORMAT statements.

4. Write a program to read the file created by either the temperature conversion program or the litres and pints conversion program. Make sure that the programs ignore the line–printer control characters, and any header and title information. This kind of problem is very common in programming (writing a program to read and possibly manipulate data created by another program).

5. Use the OPEN, REWIND, READ and WRITE statements to input a value (or values) as a character string, write this to a file, rewind the file, read in the values again, this time as real variables with blanks treated as null, then repeat with blanks as zeros.

6. Demonstrate that input and output formats are not symmetric — i.e. what goes in does not necessarily come out.

7. Can you suggest why Fortran treats blanks as null rather than zero?

8. What happens at your terminal when you enter faulty data, inappropriate for the formats specified? We will look at how we address this problem in chapter 21.

14
Functions

I can call spirits from the vasty deep.
Why so can I, or so can any man; but will they come
when you do call for them?

William Shakespeare, *King Henry IV, part 1*

Aims

The aims of this chapter are:–

- to consider some of the reasons for the inclusion of functions in a programming language
- to introduce with examples some of the predefined functions available in Fortran 90
- to introduce the concept of a user defined function
- to introduce the concept of a recursive function
- to look at scope rules in Fortran 90 for variables and functions
    - to look at global user defined functions
    - to look at internal user defined functions

14 Functions

The role of functions in a programming language and in the problem solving process is considerable and includes:–

- allowing us to refer to an action using a meaningful name, e.g. SINE(X) a very concrete use of abstraction;
- they provide a mechanism that allows us to break a problem down into parts, giving us the opportunity to structure our problem solution;
- they provide us with the ability to concentrate on one part of a problem at a time and ignore the others;
- they allow us to avoid the replication of the same or very similar sections of code when solving the same or similar sub-problem. This has the secondary effect of reducing the memory requirements of the final program;
- they allow us to build up a library of functions or modules for solving particular sub-problems, both saving considerable development time and increasing our effectiveness and productivity.

Some of the underlying attributes of functions are

- they take parameters or arguments;
- the parameter can be an expression;
- a function will normally return a value and the value returned is normally dependent on the parameter(s);
- they can sometimes take arguments of a variety of types;

Most languages provide both a range of predefined functions and the facility to define our own. We will look at the predefined functions first.

14.1 An Introduction to Predefined Functions and Their Use

Fortran provides over a hundred intrinsic functions and subroutines. A subroutine can be regarded as a variation on a function for the purposes of this chapter. They are covered in more depth in a later chapter. They are used in a straightforward way. If we take the common trigonometric functions, sine, cosine and tangent, the appropriate values may be calculated quite simply by

```
X=SIN(Y)
Z=COS(Y)
A=TAN(Y)
```

This is in rather the same way that we might say that X is a function of Y, or X is sine Y. Note that the argument, Y, is in *radians* not *degrees*.

14.1.1 Example 1: Simple function usage

A complete example is given below.

```
PROGRAM C14_Example_1
REAL :: I
   PRINT *,' Type in an angle (in radians)'
   READ *,I
   PRINT *,' Sine of ', I ,' = ',SIN(I)
END PROGRAM C14_Example_1
```

These functions are called *intrinsic functions*. A selection is given here:–

Function	Action	Example
INT	conversion to integer	J=INT(X)
REAL	conversion to real	X=REAL(J)
ABS	absolute value	X=ABS(X)
MOD	remaindering	K=MOD(I,J)
	remainder when I divided by J	
SQRT	square root	X=SQRT(Y)
EXP	exponentiation	Y=EXP(X)
LOG	natural logarithm	X=LOG(Y)
LOG10	common logarithm	X=LOG10(Y)
SIN	sine	X=SIN(Y)
COS	cosine	X=COS(Y)
TAN	tangent	X=TAN(Y)
ASIN	arcsine	Y=ASIN(X)
ACOS	arccosine	Y=ACOS(X)
ATAN	arctangent	Y=ATAN(X)
ATAN2	arctangent(a/b)	Y=ATAN2(A,B)

A complete list is given in Appendix C.

14.2 Generic Functions

All but four of the intrinsic functions and procedures are generic, i.e. they can be called with arguments of one of a number of kind types.

14.2.1 Example 2: The ABS Generic function

The following short program illustrates this with the ABS intrinsic function.

Functions

```
PROGRAM C14_Example_2
COMPLEX :: C=(1,1)
REAL    :: R=10.9
INTEGER :: I=-27
   PRINT *,ABS(C)
   PRINT *,ABS(R)
   PRINT *,ABS(I)
END PROGRAM C14_Example_2
```

The four non-generic functions are LGE, LGT, LLE and LLT – the lexical character comparison functions.

Type this program in and run it on the system you use.

It is now possible with Fortran 90 for the arguments to the intrinsic functions to be arrays. It is convenient to categorise the functions into either elemental or transformational, depending on the action performed on the array elements.

14.3 Elemental Functions

These functions work with both scalar and array arguments, i.e. with arguments that are either single or multiple valued.

14.3.1 Example 3: Elemental Function Use

Taking the earlier example with the evaluation of sine as a basis we have:–

```
PROGRAM C14_Example_3
REAL , DIMENSION(5) :: I (/1.0,2.0,3.0,4.0,5.0/)
   PRINT *,' Sine of ', I ,' = ',SIN(I)
END PROGRAM C14_Example_3
```

In the above example the sine function of each element of the array I is calculated and printed.

14.4 Transformational Functions

Transformational functions are those whose arguments are arrays, and work on these arrays to transform them in some way.

14.4.1 Example 4: Simple Transformational Use

To highlight the difference between an element by element function and a transformational function consider the following examples.

```
PROGRAM C14_Example_4
REAL , DIMENSION(5) :: I (/1.0,2.0,3.0,4.0,5.0/)
! Elemental function
```

```
   PRINT *,' Sine of ', I ,' = ',SIN(I)
! Transformational function
   PRINT *,' Sum of ', I ,' = ',SUM(I)
END PROGRAM C14_Example_4
```

The SUM function adds each element of the array and returns the SUM as a scalar, i.e. the result is single valued and not an array.

14.4.2   Example 5: Intrinsic DOT_PRODUCT use

The following program uses the transformational function DOT_PRODUCT:–

```
PROGRAM C14_Example_5
REAL , DIMENSION(5) :: I = (/1.0,2.0,3.0,4.0,5.0/)
   PRINT *,' Dot product of I with I is ',DOT_PRODUCT(I,I)
END PROGRAM C14_Example_5
```

Try typing these examples in and running them to highlight the differences between elemental and transformational functions.

14.5   Notes on Function Usage

- You should not use variables which have the same name as the intrinsic functions, e.g. what does SIN(X) mean when you have declared SIN to be a real array?

- when a function has multiple arguments care must be taken to ensure that the arguments are in the correct position and of the appropriate kind type;

- You may also replace arguments for functions by expressions, e.g.

   X = LOG(2.0)

or

   X = LOG(ABS(Y))

or

   X = LOG(ABS(Y)+Z/2.0)

14.6   Example 6: Easter

This example uses only one function, the MOD (or modulus). It is used several times, helping to emphasise the usefulness of a convenient, easily referenced function. The program calculates the date of Easter for a given year. It is derived from an algorithm by Knuth, who also gives a fuller discussion of its importance of its algorithm. He concludes that the calculation of Easter was a key factor in keeping arithmetic alive during the Middle Ages in Europe. Note, that determination of the Eastern churches' Easter requires a different algorithm.

```
PROGRAM Easter
INTEGER :: Year, Metcyc, Century, Error1, Error2, Day
INTEGER :: Epact, Luna
! A program to calculate the date of Easter
   PRINT *,' Input the year for which Easter'
   PRINT *,' is to be calculated'
   PRINT *,' enter the whole year, e.g. 1978 '
   READ *,Year
! calculating the year in the 19 year metonic cycle-metcyc
   Metcyc = MOD(Year,19)+1
   IF(Year <= 1582)THEN
      Day = (5*Year)/4
      Epact = MOD(11*Metcyc-4,30)+1
   ELSE
!          calculating the Century-century
      Century = (Year/100)+1
!          accounting for arithmetic inaccuracies
!          ignores leap years etc.
      Error1 = (3*Century/4)-12
      Error2 = ((8*Century+5)/25)-5
!          locating Sunday
      Day = (5*Year/4)-Error1-10
!          locating the epact(full moon)
      Epact = MOD(11 * Metcyc + 20 + Error2 - Error1,30)
      IF(Epact <= 0) THEN
         Epact = 30 + Epact
      ENDIF
      IF((Epact == 25 .AND. Metcyc > 11) .or. Epact == 24)THEN
         Epact = Epact+1
      ENDIF
   ENDIF
!       finding the full moon
   Luna= 44 - epact
   IF (Luna < 21) THEN
      Luna = Luna+30
   ENDIF
!       locating Easter Sunday
   Luna = Luna+7-(MOD(Day+Luna,7))
!       locating the correct month
```

```
   IF(Luna > 31)THEN
      Luna = Luna - 31
      PRINT *,' for the year ',YEAR,
      PRINT *,' Easter falls on April ',Luna
   ELSE
      PRINT *,' for the year ',YEAR,
      PRINT *,' Easter falls on march ',Luna
   ENDIF
END PROGRAM Easter
```

We have introduced a new statement here, the IF THEN ENDIF and a variant the IF THEN ELSE ENDIF. A more complete coverage is given in the later chapter on control structures. The main thing of interest is that the normal sequential flow from top to bottom can be varied. In the following case:–

IF (expression) THEN
  block of statements
ENDIF

if the expression is true the block of statements between the IF THEN and the ENDIF is executed. If the expression is false then this block is skipped, and execution proceeds with the statements immediately after the ENDIF.

In the following case:–

IF (expression) THEN
  block 1
ELSE
  block 2
ENDIF

if the expression is true block 1 is executed and block 2 is skipped. If the expression is false then block 2 is executed and block 1 is skipped. Execution then proceeds normally with the statement immediately after the ENDIF.

As well as noting the use of the MOD generic function in this program, it is also worth noting the structure of the decisions. They are *nested*, rather like the nested DO loops we met earlier.

14.7 Complete List of Predefined Functions

Due to the large number of predefined functions it is useful to classify them, and the following is one classification.

14.7.1 Inquiry Functions

These functions return information about their arguments. They can be be further sub-classified into BIT, CHARACTER, NUMERIC, ARRAY, POINTER, ARGUMENT PRESENCE

BIT	BIT_SIZE
Character	LEN
Numeric	DIGITS, EPSILON, EXPONENT, FRACTION, HUGE, KIND, MAXEXPONENT, MINEXPONENT, NEAREST, PRECISION, RADIX, RANGE, RRSPACING, SCALE, SET_EXPONENT, SELECTED_INT_KIND, SELECTED_REAL_KIND, SPACING, TINY
Array	ALLOCATED, LBOUND, SHAPE, SIZE, UBOUND,
Pointer	ASSOCIATED
Argument Presence	PRESENT

14.7.2 Transfer and Conversion Functions

These functions convert data from one type and kind type to another type and kind type. Most of them are by necessity generic.

Transfer and Conversion	ACHAR, AIMAG, AINT, ANINT, CHAR, CMPLX, CONJG, DBLE, IACHAR, IBITS, ICHAR, INT, LOGICAL, NINT, REAL, TRANSFER

14.7.3 Computational Functions

These functions actually carry out a computation of some sort and return the result of that computation.

Numeric	ABS, ACOS, ASIN, ATAN, ATAN2, CEILING, COS, COSH, DIM, DOT_PRODUCT, DPROD, EXP, FLOOR, LOG, LOG10, MATMUL, MAX, MIN, MOD, MODULO, SIGN, SIN, SINH, SQRT, TAN, TANH
Character	ADJUSTL, ADJUSTR, INDEX, LEN_TRIM, LGE, LGT, LLE, LLT, REPEAT, SCAN, TRIM, VERIFY
Bit	BTEST, IAND, IBCLR, IBSET, IEOR, IOR, ISHFT, ISHFTC, NOT

14.7.4 Array Functions

Reduction	ALL, ANY, COUNT, MAXVAL, MINVAL, PRODUCT, SUM
Construction	MERGE, PACK, SPREAD, UNPACK
Reshape	RESHAPE
Manipulation	CSHIFT, EOSHIFT, TRANSPOSE
Location	MAXLOC, MINLOC

14.7.5 Pre-Defined Subroutines

Date and Time	DATE_AND_TIME, SYSTEM_CLOCK
Random Number	RANDOM_NUMBER, RANDOM_SEED
Other	MVBITS

An alphabetical list of all intrinsic functions and procedures is given in Appendix C. This list provides the following information:–

- function name
- description
- argument name and type
- result type
- classification
- examples of use

This appendix should be consulted for a more complete and thorough understanding of intrinsic functions and their use in Fortran 90.

14.8 Supplying your own functions

There are two stages here, firstly to define the function and secondly to reference or use it. Consider the calculation of the greatest common divisor of two integers.

14.8.1 Example 7: Simple User Defined Function

The following defines a function to achieve this:–

```
INTEGER FUNCTION GCD(A,B)
IMPLICIT NONE
INTEGER , INTENT(IN) :: A,B
INTEGER :: Temp
   IF (A < B) THEN
      Temp=A
   ELSE
      Temp=B
   ENDIF
   DO WHILE ((MOD(A,Temp) /= 0) .OR. (MOD(B,Temp) /=0))
      Temp=Temp-1
   END DO
   GCD=Temp
END FUNCTION GCD
```

Functions

To use this function, you reference or call it with a form like:—

```
PROGRAM GCDExample
IMPLICIT NONE
INTEGER :: I,J,Result
INTEGER :: GCD
INTEGER :: T,TR,TMax
   PRINT *,' Type in two integers'
   READ *,I,J
   CALL SYSTEM_CLOCK(T,TR,TMax)
   PRINT *,T
   Result=GCD(I,J)
   CALL SYSTEM_CLOCK(T,TR,TMax)
   PRINT *,T
   PRINT *,' GCD is ',Result
END PROGRAM GCDExample
```

The first line of the function

```
INTEGER FUNCTION GCD(A,B)
```

has a number of things of interest:—

- firstly the function has a type, and in this case the function is of type INTEGER, i.e. it will return an integer value;

- the function has a name, in this case GGD

- the function takes arguments or parameters, in this case A and B.

The structure of the rest of function is the same as that of a program, i.e. we have declarations, followed by the executable part. This is because both a program and a function can be regarded as a so called *program unit*. We will look into this more fully in later chapters.

In the declaration we also have a new attribute for the INTEGER declaration. The two parameters A and B are of type integer, and the INTENT(IN) attribute means that these parameters will NOT be altered by the function.

The value calculated is returned through the function name somewhere in the body of the executable part of the function. In this case GCD appears on the left hand side of an arithmetic assignment statement at the bottom of the function. The end of the function is signified in the same way as the end of a program:—

```
END FUNCTION GCD
```

We then have the program which actually uses the function GCD. In the program the function is called or invoked with I and J as arguments. The variables are called A and B in the function, and references to A and B in the function will use the values that I and J have respectively in the main program. We will look into the whole area of argument association in much greater depth in later chapters.

Note also here a new control statement, the DO WHILE ENDDO. In the following case:–

DO WHILE (expression)
  block of statements
ENDDO

the block of statements between the DO WHILE and the ENDDO is executed whilst the expression is true. There is a more complete coverage in the chapter on control structures.

Note also the use of one of the predefined subroutines SYSTEM_CLOCK. Subroutines are invoked by the use of the CALL statement, rather than the function name. The first argument returns a value based on an implementation specific real time clock. We will look into subroutines in more depth in a later chapter.

We have two options here regarding compilation. Firstly to make the function and the program into one file, and invoke the compiler once. Secondly to make the function and program into separate files, and invoke the compiler twice, once for each file. With large programs comprising one program and several functions it is probably worthwhile to keep the component parts in different files and compile individually, whilst if a simple program and one function keeping things together in one file.

Try this program out on the system you work with, and look at the timing information provided.

14.9 An Introduction to the Scope of Variables and Local Variables

One of the major strengths of Fortran is the ability to work on parts of a problem at a time. This is achieved by the use of *program units* (a main program, one or more functions and one or more subroutines) to solve discrete sub-problems. Interaction between them is limited and can be isolated, for example, to the arguments of the function. Thus variables in the main program can have the same name as variables in the function and they are completely separate variables, even though they have the same name. Thus we have the concept of a local variable in a program unit. We will look into this area again after a coverage of recursion and very thoroughly after the coverage of subroutines and modules.

In the example above I, J, Result, T, TR and TMax are local to the main program. The declaration of GCD is to tell the compiler that it is an integer, and in this case is is an external function.

A and B in the function GCD do not exist in any real sense, rather that they will be replaced by the actual variable values from the calling routine, in this case by whatever values I and J have. Temp is local to GCD.

14.10 Recursive Functions

There is an additional form of the function header that must be used when the function is recursive. Recursion means the breaking down of a problem into a simpler but identical sub problem. The concept is best explained with reference to an actual example. Consider the evaluation of a factorial, e.g. 5!. From simple mathematics we know that the following is true

5!=5*4!

4!=4*3!

3!=3*2!

2!=2*1!

1!=1

and thus 5! = 5*4*3*2*1 or 120.

14.10.1 Example 8: Recursive Factorial Evaluation

Let us look at a program with recursive function to solve the evaluation of factorials.

```
PROGRAM FactorialEvaluation
IMPLICIT NONE
INTEGER :: I, F, Factorial
   PRINT *,' Type in the number, integer only'
   READ *,I
   DO WHILE(I<0)
      PRINT *,' Factorial only defined for positive integers'
      PRINT *,' Re-input'
      READ *,I
   END DO
   F=Factorial(I)
   PRINT *,' Answer is', F
END PROGRAM FactorialEvaluation

RECURSIVE INTEGER FUNCTION Factorial(I) RESULT(Answer)
IMPLICIT NONE
INTEGER , INTENT(IN):: I
   IF (I==0) THEN
      Answer=1
   ELSE
      Answer=I*Factorial(I-1)
   END IF
END FUNCTION Factorial
```

Well what additional information is there? Firstly we have an additional attribute on the function header that declares the function to be recursive. Secondly we must return the result in a variable, in this case Answer. Let us look now at what happens when we compile and run the whole program (both function and main program). If we type in the number 5 the following will happen:–

- the function is first invoked with argument 5. The ELSE block is then taken and the function is invoked again;

- the function now exists a second time with argument 4. The ELSE block is then taken and the function is invoked again;

- the function now exists a third time with argument 3. The ELSE block is then taken and the function is invoked again;

- the function now exists a fourth time with argument 2. The ELSE block is then taken and the function is invoked again;

- the function now exists a fifth time with argument 1. The ELSE block is then taken and the function is invoked again;

- the function now exists a sixth time with argument 0. The IF BLOCK is executed and Answer=1. This invocation ends and we return to the previous level, with Answer=1*1.

- We return to the previous invocation and now Answer=2*1.

- We return to the previous invocation and now Answer=3*2.

- We return to the previous invocation and now Answer=4*6.

- We return to the previous invocation and now Answer=5*24

The function now terminates and we return to the main program or calling routine. The answer 120 is the printed out.

Add a PRINT *,I statement to the function after the last declaration and type the program in and run it. Try it out with 5 as the input value to verify the above statements.

Recursion is a very powerful tool in programming, and remarkably simple solutions are possible to quite complex problems using recursive techniques. We will look at recursion in much more depth in the later chapters on dynamic data types, and subroutines and modules.

14.11 Example 9: Recursive version of GCD

The following is another example of the earlier GCD function but with the algorithm in the function replaced with an alternate recursive solution.

```
PROGRAM GCDExample
IMPLICIT NONE
```

```
INTEGER  ::  I,J,Result
INTEGER  ::  GCD
INTEGER  ::  T,TR,TMax
   PRINT *,' Type in two integers'
   READ *,I,J
   CALL SYSTEM_CLOCK(T,TR,TMax)
   PRINT *,T
   Result=GCD(I,J)
   CALL SYSTEM_CLOCK(T,TR,TMax)
   PRINT *,T
   PRINT *,' GCD is ',Result
END PROGRAM GCDExample

RECURSIVE INTEGER FUNCTION GCD(I,J) RESULT(Answer)
IMPLICIT NONE
INTEGER , INTENT(IN) :: I,J
   IF (J==0) THEN
      Answer=I
   ELSE
      Answer=GCD(J,MOD(I,J))
   ENDIF
END FUNCTION GCD
```

Try this program out on the system you work with, and look at the timing information provided, and compare the timings with the previous example. The algorithm is a much more efficient algorithm than in the original example, and hence should be much faster. On one system there was a twenty fold decrease in execution time between the two versions.

Recursion is sometimes said to be inefficient, and the following example looks at a non-recursive version of the second algorithm.

14.12 Example 10: After Removing Recursion

The following is a variant of the above, with the same algorithm, but with the recursion removed.

```
PROGRAM GCDExample
IMPLICIT NONE
INTEGER :: I,J,Result
INTEGER :: GCD
INTEGER :: T,TR,TMax
   PRINT *,' Type in two integers'
   READ *,I,J
   CALL SYSTEM_CLOCK(T,TR,TMax)
   PRINT *,T
```

```
      Result=GCD(I,J)
      CALL SYSTEM_CLOCK(T,TR,TMax)
      PRINT *,T
      PRINT *,' GCD is ',Result
END PROGRAM GCDExample

INTEGER FUNCTION GCD(I,J)
IMPLICIT NONE
INTEGER , INTENT(INOUT) :: I,J
INTEGER :: Temp
   DO WHILE (J/=0)
      Temp=MOD(I,J)
      I=J
      J=Temp
   END DO
   GCD=I
END FUNCTION GCD
```

Compare the timing information with the previous examples. Is this non-recursive version significantly faster?

14.13 Internal functions

An internal function is a more restricted and hidden form of the normal function definition.

Since the internal function is specified within a program segment, it may only be used within that segment, and cannot be referenced from any other functions or subroutines, unlike the intrinsic or other user-defined functions.

14.13.1 Example 11: Stirling's Approximation

In this example we use Stirling's approximation for large N

$$n! = \sqrt{2\pi n} \left(\frac{n}{e}\right)^n$$

and a complete program to use this internal function is given below.

```
PROGRAM Factor
IMPLICIT NONE
REAL :: Result,N,R
   PRINT *,' Type in N and R'
   READ *,N,R
!  NUMBER OF POSSIBLE COMBINATIONS THAT CAN BE FORMED WHEN
```

Functions

```
!   R OBJECTS ARE SELECTED OUT OF A GROUP OF N
!         N!/R!(N-R)!
    Result=Stirling(N)/(Stirling(R)*Stirling(N-R))
    PRINT *,Result
    PRINT *,N,R
CONTAINS
REAL FUNCTION Stirling (X)
    REAL , INTENT(IN) :: X
    REAL , PARAMETER :: PI=3.1415927, E =2.7182828
    Stirling=SQRT(2.*PI*X) * (X/E)**X
END FUNCTION Stirling
END PROGRAM Factor
```

The difference between this example and the earlier ones lies in the CONTAINS statement. The function is now an integral part of the program, and could not be used elsewhere in another function for example. This provides us with a very powerful way of information hiding, and making the construction of larger programs more secure and bug free.

14.14 Resume

Within the world of mathematics there is the concept of a pure function. This means that the function only returns a value, and has no effect on the arguments. As a general rule we recommend this as it makes the construction of larger programs more secure, and less error prone. There are problems in this area when we are interested in trapping errors, but we will look into this area in more depth later on.

There are a large number of Fortran supplied functions and subroutines (intrinsic functions) which extend the power and scope of the language. Some of these functions are of *generic* type, and can take several different types of argument. Others are restricted to a particular type of argument. Appendix C should be consulted for a fuller coverage concerning the rules that govern the use of the intrinsic functions and procedures.

When the intrinsic functions are inadequate, it is possible to write *user defined* functions. Besides expanding the scope of computation, such functions help in problem visualisation and logical subdivision, may reduce duplication, and generally help in avoiding programming errors.

In addition to separately defined user functions, internal functions may be employed. These are functions which are used within a program segment.

Although the normal exit from a user defined function is through the END, other, *abnormal*, exits may be defined through the RETURN statement.

Communication with non-recursive functions is through the function name and the function arguments. The function *must* contain a reference to the function name on the left hand side of an assignment. Results may also be returned through the argument list.

We have also covered briefly the concept of scope for a variable, local variables, and argument association. This area warrants a much fuller coverage and we will do this after we have covered subroutines and modules.

14.15 Function Syntax

The syntax of a function is:–

[function prefix] function_statement &
[RESULT (Result_name)]
[specification part]
[execution_part]
[internal sub program part]
END [FUNCTION [function name]]

and prefix is:–

[type specification] RECURSIVE

or

[RECURSIVE] type specification

and the function_statement is:–

FUNCTION function_name ([dummy argument name list])

[ ] represent optional parts to the specification.

14.16 Rules and Restrictions

The type of the function must only be specified once, either in the function statement or in a type declaration.

The names must match between the function header and END FUNCTION function name statement.

If there is a RESULT clause that name must be used as the result variable, and therefore all references to the function name are recursive calls.

The function name must be used to return a result when there is no RESULT clause.

We will look at additional rules and restrictions in later chapters.

14.17 Problems

1. Find out the action of the MOD function when one of the arguments is negative. Write your own modulus function to return only a positive remainder. Don't call it MOD!

2. Create a table which gives the sines, cosines and tangents for 0 degrees to 90 degrees in 1 degree intervals. There are a few minor catches in this question.

3. Write a function to convert an integer to a binary character representation. It should take an integer argument and return a character string that is a sequence of zeros and ones. Use the program in chapter eight as a basis for the solution.

14.18 Bibliography

Abramowitz M., Stegun I., *Handbook of Mathematical Functions*, Dover.

- This book contains a fairly comprehensive collection of numerical algorithms for many mathematical functions, of varying degrees of obscurity. It is a widely used source.

Association of Computing Machinery (ACM)

- *Collected Algorithms*, 1960–1974
- *Transactions on Mathematical Software*, 1975 –
  A good source of more specialised algorithms. Early algorithms tended to be in Algol, Fortran now predominates.

14.18.1 Recursion and Problem Solving

The following are a number of books that look at the role of recursion in problem solving and algorithms.

Hofstader D. R., *Godel, Escher, Bach – an Eternal Golden Braid*, Harvester Press.

- The book provides a stimulating coverage of the problems of paradox and contradiction in art music and mathematics using the works of Escher, Bach and Godel, and hence the title. There is a whole chapter on recursive structures and processes. The book also covers the work of Church and Turing, both of whom have made significant contributions to the theory of computing.

Kruse R.L., *Data Structures and Program Design*, Prentice Hall

- Quite a gentle introduction to the use of recursion and its role in problem solving. Good choice of case studies with explanations of solutions. Pascal is used.

Sedgewick R., *Algorithms*, Addison Wesley.

- Good source of algorithms. Well written. The GCD algorithm was taken from this source.

Wirth N., *Algorithms + Data Structures = Programs*, Prentice Hall,

- In the context of this chapter the section on recursive algorithms is a very worthwhile investment in time.

Wood D., *Paradigms and Programming in Pascal*, Computer Science Press.

- Contains a number of examples of the use of recursion in problem solving. Also provides a number of useful case studies in problem solving.

15
Control Structures

Summarising: as a slow-witted human being I have a very small head and I had better learn to live with it and to respect my limitations and give them full credit, rather than try to ignore them, for the latter vain effort will be punished by failure.
Edsger W. Dijkstra, *Structured Programming*.

Aims

The aims of this chapter are to introduce:–

- selection between various courses of action as part of the problem solution
- the concepts and statements in Fortran needed to support the above. In particular:–
    - logical expressions
    - logical operators
    - a *block* of statements
    - several *blocks* of statements
- the IF THEN ENDIF construct
- the IF THEN ELSE IF ENDIF construct
- to introduce the CASE statement with examples
- to introduce the DO loop, in three forms with examples, in particular
    - the iterative DO loop
    - the DO WHILE form
    - the DO ... IF THEN EXIT END DO or repeat until form
    - the CYCLE statement
    - the EXIT statement

15 Control Structures

When we look at the this area it is useful to gain some historical perspective concerning what control structures are available in a programming language.

At the time of the development of Fortran in the 1950's there was little theoretical work around and the control structures provided were very primitive and very closely related to the capability of the hardware.

By the time of the first standard in 1966 there was little published work regarding structured programming and control structures. The seminal work by Dahl, Dijkstra and Hoare was not published until 1972.

By the time of the second standard there was a major controversy regarding languages with poor controls structures like Fortran which essentially were limited to the GOTO statement. The facilities in the language had lead to the development and continued existence of major code suites that were unintelligible, and the pejorative term spaghetti was applied to these programs. Developing an understanding of what a program did became an almost impossible task in many cases.

Fortran missed out in 1977 of incorporating some of the more modern and intelligible control structures that had emerged as being of major use in making code more easy to understand and modify.

It was not until the latest standard that a reasonable set of control structures have emerged and become an accepted part of the language. The more inquisitive reader is urged to read at least the work by Dahl, Dijkstra and Hoare to develop some understanding of the importance of control structures and the role of structured programming. The paper by Knuth is also highly recommended and provides a very balanced coverage of the controversy of earlier times over the GOTO statement.

15.1 Selection between courses of action

In most problems you need to chose between various courses of action e.g.

- if overdrawn, then do not draw money out of the bank
- if Monday, Tuesday, Wednesday, Thursday or Friday, then go to work
- if Saturday, then go to watch Queens Park Rangers
- if Sunday, then lie in bed for another two hours

As most problems involve selection between two or more courses of action it is necessary to have the concepts to support this in a programming language. Fortran has a variety of selection mechanisms, some of which are introduced.

15.1.1 The BLOCK IF statement.

The following short example illustrates the main ideas:–

. wake up
.
. check the date and time
IF (Today = = Sunday) THEN

    .
    . lie in bed for another two hours
    .

ENDIF
    .
. get up
. make breakfast

If today is Sunday then the block of statements between the IF and the ENDIF is executed. After this block has been executed the program continues with the statements after the ENDIF. If today is not Sunday the program continues with the statements after the ENDIF immediately. This means that the statements after the ENDIF are executed whether or not the expression is true.

The general form is:–

    IF (Logical expression) THEN
        .
        Block of statements
        .
    ENDIF

The logical expression is an expression that will be either true or false, hence its name. Some examples of logical expressions are given below:–

```
(Alpha >= 10.1)
```
    Test if Alpha 10.1

```
(Balance <= 0.0)
```
    Test if overdrawn

```
(( Today == Saturday).OR.( Today == Sunday))
```
    Test if today is Saturday or Sunday

```
((Actual-Calculated) <= 1.0E-6)
```
    Test if Actual minus Calculated less than or equal to 1.0E-6

Fortran has the following relational and logical operators:–

Control Structures

Operator	Meaning	Type
==	Equal	Relational
/=	Not equal	Relational
>=	Greater than or equal	Relational
<=	Less than or equal	Relational
<	Less than	Relational
>	Greater than	Relational
.AND.	and	Logical
.OR.	or	Logical
.NOT.	not	Logical

The first six should be self-explanatory. They enable expressions or variables to be compared and tested. The last three enable the construction of quite complex comparisons, involving more than one test; in the example given earlier there was a test to see whether today was Saturday or Sunday.

Use of logical expressions and logical variables (something not mentioned so far) are covered again in a later chapter on logical data types.

The 'IF *expression* THEN *statements* ENDIF' is called a BLOCK IF construct. There is a simple extension to this provided by the ELSE statement. Consider the following example:–

```
    IF (Balance > 0.0) THEN
            . draw money out of the bank
    ELSE
            . borrow money from a friend
    ENDIF
    . Buy a round of drinks.
```

In this instance, one or other of the blocks will be executed. Then execution will continue with the statements after the ENDIF statement (in this case *buy a round*).

There is yet another extension to the BLOCK IF which allows ELSEIF statement. Consider the following example:–

```
IF (Today == Monday) THEN
   .
ELSEIF (Today == Tuesday) THEN
   .
ELSEIF (Today == Wednesday) THEN
   .
```

```
    ELSEIF  (Today  ==  Thursday)  THEN
         .
    ELSEIF  (Today  ==  Friday)  THEN
         .
    ELSEIF  (Today  ==  Saturday)  THEN
         .
    ELSEIF  (Today  ==  Sunday)  THEN
         .
    ELSE
       there has been an error. The variable Today has
       taken on an illegal value.
    ENDIF
```

Note that, as soon as one of the logical expressions is true, the rest of the test is skipped, and execution continues with the statements after the ENDIF. This implies that a construction like:–

```
    IF(I < 2)THEN
       ...
    ELSEIF(I < 1)THEN
       ...
    ELSE
       ...
    ENDIF
```

is inappropriate. If I is less than 2, the latter condition will never be tested. The ELSE statement has been used here to aid in trapping errors or exceptions. This is recommended practice. A very common error in programming is to assume that the data is in certain well-specified ranges. The program then fails when the data goes outside this range. It makes no sense to have a day other than Monday, Tuesday, Wednesday, Thursday, Friday, Saturday or Sunday.

15.1.2 Example 1: Quadratic Roots

This program is straightforward, with a simple structure. The roots of the quadratic are either real, equal and real, or complex depending on the magnitude of the term B ** 2 – 4 * A * C. The program tests for this term being greater than and less than zero, it assumes that the only other case is equality to zero (from the mechanics of a computer, floating point equality is rare, but, we are safe in this instance).

```
PROGRAM Qroots
IMPLICIT NONE
REAL :: A , B , C , Term , A2 , Root1 , Root2
!
!     a b and c are the coefficients of the terms
!     a*x**2+b*x+c
```

```
!    find the roots of the quadratic, root1 and root2
!
     PRINT*,' GIVE THE COEFFICIENTS A, B AND C'
     READ*,A,B,C
     Term = B*B - 4.*A*C
     A2 = A*2.
!  if term < 0, roots are complex
!  if term = 0, roots are equal
!  if term > 0, roots are real and different
     IF(Term < 0.0)THEN
        PRINT*,' ROOTS ARE COMPLEX'
     ELSEIF(Term > 0.0)THEN
        Term = SQRT(Term)
        Root1 = (-B+Term)/A2
        Root2 = (-B-Term)/A2
        PRINT*,' ROOTS ARE ',Root1,' AND ',Root2
     ELSE
        Root1 = -B/A2
        PRINT*,' ROOTS ARE EQUAL, AT ',Root1
     ENDIF
END PROGRAM Qroots
```

15.1.3 Note

Given the understanding you now have about real arithmetic and finite precision will the ELSE block above ever be executed?

15.1.4 Example 2: Date calculation

This next example is also straightforward. It demonstrates that, even if the conditions on the IF statement are involved, the overall structure is easy to determine. The comments and the names given to variables should make the program self-explanatory. Note the use of integer division to identify leap years.

```
PROGRAM Date
IMPLICIT NONE
INTEGER :: Year , N , Month , Day , T
!
! calculates day and month from year and day-within-year
!   t is an offset to account for leap years. Note that the first
! criteria is division by 4, but that centuries are only
! leap years if divisible by 400, not 100 (4 * 25) alone.
!
     PRINT*,' year, followed by day within year'
```

```
      READ*,Year,N
!     checking for leap years
      IF ((Year/4)*4 == Year ) THEN
         T=1
         IF ((Year/400)*400 == Year ) THEN
            T=1
         ELSEIF ((Year/100)*100 == Year) THEN
            T=0
         ENDIF
      ELSE
         T=0
      ENDIF
!     accounting for February
      IF(N > (59+T))THEN
         Day=N+2-T
      ELSE
         Day=N
      ENDIF
      Month=(Day+91)*100/3055
      Day=(Day+91)-(Month*3055)/100
      Month=Month-2
      PRINT*,' CALENDAR DATE IS ', Day , Month , Year
END PROGRAM Date
```

15.1.5 The CASE Statement

The CASE statement provides a very clear and expressive selection mechanism between two or more course of action. Strictly speaking it could be constructed from the IF THE ELSE IF ENDIF statement, but with considerable loss of clarity. Remember that programs have to read and understood by both humans and compilers!

15.1.6 Example 3: Simple calculator

```
PROGRAM Case01
IMPLICIT NONE
!
! Simple case statement example
!
INTEGER :: I,J,K
CHARACTER :: Operator
   DO
      PRINT *,' Type in two integers'
      READ *, I,J
```

Control Structures

```
      PRINT *,' Type in operator'
      READ '(A)',Operator
      Calculator : &
      SELECT CASE (Operator)
         CASE ('+') Calculator
            K=I+J
            PRINT *,' Sum of numbers is ',K
         CASE ('-') Calculator
            K=I-J
            PRINT *,' Difference is ',K
         CASE ('/') Calculator
            K=I/J
            PRINT *,' Division is ',K
         CASE ('*') Calculator
            K=I*J
            PRINT *,' Multiplication is ',K
      CASE DEFAULT Calculator
         EXIT
      END SELECT Calculator
   END DO
END PROGRAM Case01
```

The user is prompted to type in two integers and the operation that they would like carried out on those two integers. The CASE statement then ensures that the appropriate arithmetic operation is carried out. The program terminates when the user types in any other character than +, -, * or /.

The CASE DEFAULT options introduces the EXIT statement. This statement is used in conjunction with the DO statement. When this statement is executed control passes to the statement immediately after the matching END DO statement. In the example above the program terminates, as there are no executable statements after the END DO.

15.1.7 Example 4: Counting Vowels, Consonants, etc

This example is more complex, but again is quite easy to understand. The user types in a line of text and the program produces a summary of the frequency of the characters typed in.

```
PROGRAM Case02
IMPLICIT NONE
!
! Simple counting of vowels, consonants,
! digits, blanks and the rest
!
INTEGER :: Vowels=0 , Consonants=0, Digits=0, Blank=0, Other=0, I
CHARACTER :: Letter
```

```
CHARACTER (LEN=80) :: Line
   READ '(A)', Line
   DO I=1,80
      Letter=Line(I:I)  ! extract one character at position I
      Count : &
      SELECT CASE (Letter)
         CASE ('A','E','I','O','U', &
               'a','e','i','o','u') Count
            Vowels=Vowels + 1
         CASE ('B','C','D','F','G','H', &
               'J','K','L','M','N','P', &
               'Q','R','S','T','V','W', &
               'X','Y','Z',             &
               'b','c','d','f','g','h', &
               'j','k','l','m','n','p', &
               'q','r','s','t','v','w', &
               'x','y','z')Count
            Consonants=Consonants + 1
         CASE ('1','2','3','4','5','6','7','8','9','0') Count
            Digits=Digits + 1
         CASE (' ') Count
            Blank=Blank + 1
         CASE DEFAULT Count
            Other=Other+1
      END SELECT Count
   END DO
   PRINT *, ' Vowels =  ', Vowels
   PRINT *, ' Consonants =  ', Consonants
   PRINT *, ' Digits = ', Digits
   PRINT *, ' Blanks = ',Blank
   PRINT *, ' Other characters = ', Other
END PROGRAM Case02
```

15.2 The three forms of the DO statement.

You have already been introduced in the chapters on arrays to the iterative form of the DO loop, i.e.

DO Variable = Start, End, Increment

   block of statements

END DO

A complete coverage of this form is given in the three chapters on arrays.

Control Structures

There are two additional forms of the block DO that complete our requirements and these are:–

- DO WHILE (Logical Expression) ... ENDDO

and

- DO ... IF (Logical Expression) EXIT END DO

The first form is often called a WHILE loop as the block of statements executes whilst the logical expression is true, and the second form is often called a REPEAT UNTIL loop as the block of statements executes until the statement is true.

Note that the WHILE block of statements may never be executed, and the REPEAT UNTIL block will always be executed at least once.

15.2.1 Example 5: Sentinel Usage

The following example shows a complete program using this construct.

```
PROGRAM Find
IMPLICIT NONE
! this program picks up the first occurrence
! of a number in a list.
! a sentinel is used, and the array is 1 more
! than the max size of the list.
INTEGER , DIMENSION(101) :: A
INTEGER :: Mark
INTEGER :: End,I
   READ (UNIT=1,FMT=*) Mark
   READ (UNIT=1,FMT=*) End
   READ(UNIT=1,FMT=*)  (A(I),I=1,End)
   I=1
   A(End+1)= Mark
   DO WHILE(Mark /= A(I))
      I=I+1
   END DO
   IF(I == (END+1)) THEN
      PRINT*,' ITEM NOT IN LIST'
   ELSE
      PRINT*,' ITEM IS AT POSITION ',I
   ENDIF
END PROGRAM Find
```

The *repeat until* construct is written in Fortran as:–

```
DO
   ...
   ...
   IF (Logical Expression) EXIT
END DO
```

There are problems in most disciplines that require a numerical solution. The two main reasons for this are that either the problem can only be solved numerically, or that an analytic solution involves too much work. Solutions to this type of problem often require the use of the *repeat until* construct. The problem will typically require the repetition of a calculation until the answers from successive evaluations differ by some small amount, decided generally by the nature of the problem. Here is a program extract to illustrate this:–

```
REAL , PARAMETER :: TOL=1.0E-6
   .
DO
   ...
   CHANGE=
   ...
   IF (CHANGE <= TOL) EXIT
END DO
```

The value of the tolerance is set here to 1.0E–6. Note again the use of the EXIT statement. The DO END DO block is terminated and control passes to the statement immediately after the matching END DO.

15.2.2 CYCLE and EXIT

These two statements are used in conjunction with the block DO statement. You have seen examples above of the use of the EXIT statement to terminate the block DO, and pass control to the statement immediately after the corresponding END DO statement.

The CYCLE statement can appear anywhere in a block DO and will immediately pass control to the start of the block DO. Examples of CYCLE and EXIT are given in later chapters.

15.2.3 Example 6: e**x Evaluation

The function ETOX illustrates one use of the *repeat until* construct. The function evaluates e**x. This may be written as:–

$$1 + x/1! + x^2/2! + x^3/3! \ ...$$

or

Control Structures

$$1 + \sum_{n=1}^{\infty} \frac{x^{n-1}}{(n-1)!} \frac{x}{n}$$

Every succeeding term is just the previous term multiplied by x/n. At some point the term x/n becomes very small, so that it is not sensibly different from zero, and successive terms add little to the value. The function therefore repeats the loop until x/n is smaller than the tolerance. The number of evaluations is not known beforehand, since this is dependent on x.

```
REAL FUNCTION Etox(X)
IMPLICIT NONE
REAL :: Term , X
INTEGER :: Nterm
REAL , PARAMETER ::Tol = 1.0E-6
   Etox=1.0
   Term=1.0
   Nterm=0
   DO
      Nterm = Nterm +1
      Term =( X / Nterm) * Term
      Etox = Etox + Term
      IF(Term <= Tol)EXIT
   END DO
END FUNCTION Etox
```

Both types of loop are combined in this last example. The algorithm employed here finds the zero of a function. Essentially, it finds an interval in which the zero must lie; the evaluations on either side are of different sign. The *while loop* ensures that the evaluations are of different sign, by exploiting the knowledge that the incident wave height must be greater than the reformed wave height (to give the lower bound). The upper bound is found by experiment, making the interval bigger and bigger. Once the interval is found, its mean is used as a new potential bound. The zero must lie on one side or the other; in this fashion, the interval containing the zero becomes smaller and smaller, until it lies within some tolerance. This approach is rather plodding and unexciting, but is suitable for a wide range of problems.

15.2.4 Example 7: Wave Breaking on an Offshore Reef

This example is drawn from a situation where a wave breaks on an offshore reef or sand bar, and then reforms in the near-shore zone before breaking again on the coast. It is easier to observe the heights of the reformed waves reaching the coast than those incident to the terrace edge.

```
PROGRAM Break
IMPLICIT NONE
REAL :: Hi , Hr , Hlow , High , Half , Xl , Xh , Xm , D
```

```
REAL , PARAMETER :: Tol=1.0E-6
! problem - find hi from expression given in function f
! F=A*(1.0-0.8*EXP(-0.6*C/A))-B
! HI IS INCIDENT WAVE HEIGHT            (C)
! HR IS REFORMED WAVE HEIGHT            (B)
! D IS WATER DEPTH AT TERRACE EDGE      (A)
   PRINT*,' Give reformed wave height, and water depth'
   READ*,Hr,d
!
! for Hlow- let Hlow=hr
! for high- let high=Hlow*2.0
!
! check that signs of function results are different
!
   Hlow = Hr
   High = Hlow*2.0
   Xl = F( Hlow, Hr, D)
   Xh = F( High, Hr, D)
!
   DO WHILE ((XL*XH)  >= 0.0)
      HIGH = HIGH*2.0
      XH  = F(HIGH,HR,D)
   END DO
!
   DO
      HALF=(HLOW+HIGH)*0.5
      XM=F(HALF,HR,D)
      IF((XL*XM)  < 0.0)THEN
         XH=XM
         HIGH=HALF
      ELSE
         XL=XM
         HLOW=HALF
      ENDIF
      IF(ABS(HIGH-HLOW)<= TOL)EXIT
   END DO
   PRINT*,' Incident Wave Height Lies Between'
   PRINT*,Hlow,' and ',High,' metres'
CONTAINS
REAL FUNCTION F(A,B,C)
IMPLICIT NONE
REAL , INTENT (IN) :: A
```

Control Structures

```
REAL , INTENT (IN) :: B
REAL , INTENT (IN) :: C
   F=A*(1.0-0.8*EXP(-0.6*C/A))-B
END FUNCTION F
END PROGRAM Break
```

15.3 Summary

You have been introduced in this chapter to several control structures and these include:–

- the *block if*
- the *if then else if*
- the *case* construct
- the block *do* in three forms:–
  - the *iterative do* or *do variable=start,end,increment ... end do*
  - the *while* construct, or *do while ... end do*
  - the *repeat until* construct, or *do ... if then exit end do*
- the *cycle* and *exit* statements, which can be used with *do* statement in all three forms:–
  - the *do variable = start,end,increment ... end do*
  - the *while* construct, or *do while ... end do*
  - the *repeat until* construct, or *do ... if then exit end do*

These constructs are sufficient to solve a wide class of problems. There are other control statements available in Fortran especially those inherited from Fortran 66 and Fortran 77, but the ones that are covered here are the preferred ones. We will look in chapter 28 at one more control statement, the so called GOTO statement, with recommendations as to where its use is appropriate.

The above are more formally described below:–

CASE

```
SELECT CASE ( case variable )
   [ CASE case selector
      [executable construct ] ... ] ...
   [ CASE DEFAULT
   [executable construct ]
END SELECT
```

DO

```
DO [ label ]
   [executable construct ] ...
do termination

DO [ label ] [ , ] loop variable = initial value , final value
, [ increment ]
   [executable construct ] ...
do termination

DO [ label ] [ , ] WHILE (scalar logical expression )
   [executable construct ] ...
do termination
```

IF

```
IF ( scalar logical expression ) THEN
   [executable construct ] ...
[ ELSE IF ( scalar logical expression   THEN
   [executable construct ] ... ] ...]
[ ELSE
      [executable construct ] ...]
END IF
```

WHERE

```
WHERE ( array logical expression )
   array assignment block
ELSEWHERE
   array assignment block
END WHERE
```

15.4 Problems

1. Rewrite the program for period of a pendulum. The new program should print out the length of the pendulum, and period for lengths of the pendulum from 0 to 100 cm in steps of 0.5 cm. The program should incorporate a function for the evaluation of the period.

2. Using functions, do the following:–

- Evaluate n! from n=0 to n=10
- Calculate 76 factorial.
- Now calculate $(x^{**}n)/n!$, with x=13.2 and n=20.

Control Structures

- Now do it another way.

3. The program BREAK is taken from a real example. In the particular problem, the reformed wave height was 1 metre, and the water depth at the reef edge was 2 metres. What was the incident wave height? Rather than using an absolute value for the tolerance, it might be more realistic to use some value related to the reformed wave height. These heights are unlikely to be reported to better than about 5 per cent accuracy. Wave energy may be taken as proportional to wave height squared for this example. What is the reduction in wave energy as a result of breaking on the reef or bar, for this particular case.

4. What is the effect of using INT on negative real numbers? Write a program to demonstrate this.

5. How would you find the nearest integer to a real number? Now do it another way. Write a program to illustrate both methods. Make sure you test it for negative as well as positive values.

6. The function ETOX has been given in this chapter. The standard Fortran function EXP does the same job. Do they give the same answers? Curiously the Fortran standard does not specify how a *standard* function should be evaluated, or even how accurate it should be.

The physical world has many examples where processes require some threshold to be overcome before they begin operation: critical mass in nuclear reactions, a given slope to be exceeded before friction is overcome, and so on. Unfortunately, most of these sorts of calculations become rather complex and not really appropriate here. The following problem tries to restrict the range of calculation, whilst illustrating the possibilities of decision making.

7. If a cubic equation is expressed as

$$z^3 + a_2 z^2 + a_1 z + a_0 = 0$$

and we let

$$q = a_1/3 - (a_2 * a_2)/9$$

and

$$r = (a_1 a_2 - 3 a_0)/6 - (a_2 a_2 a_2)/27$$

we can determine the nature of the roots as follows:

$q^3 + r^2 > 0$; one real root and a pair of complex;
$q^3 + r^2 = 0$; all roots real, and at least two equal;
$q^3 + r^2 < 0$; all roots real;

Incorporate this into a suitable program, to determine the nature of the roots of a cubic from suitable input.

8. The form of breaking waves on beaches is a continuum, but for convenience we commonly recognise three major types: surging, plunging and spilling. These may be classified empirically by reference to the wave period, T (seconds), the breaker wave height, H_b (metres), and the beach slope, m. These three variables are combined into a single parameter, B, where

$$B = H_b/(gmT^2)$$

g is the gravitational constant (981 cm sec^{-2}). If B is less than .003, the breakers are surging; if B is greater than 0.068, they are spilling, and between these values, plunging breakers are observed.

(i) On the east coast of New Zealand, the normal pattern of waves is swell waves, with wave heights of 1 to 2 metres, and wave periods of 10 to 15 seconds. During storms, the wave period is generally shorter, say 6 to 8 seconds, and the wave heights higher, 3 to 5 metres. The beach slope may be taken as about 0.1. What changes occur in breaker characteristics as a storm builds up?

(ii) Similarly, many beaches have a concave profile. The lower beach generally has a very low slope, say less than 1 degree (m=0.018), but towards the high tide mark, the slope increases dramatically, to say 10 degrees or more (m=0.18). What changes in wave type will be observed as the tide comes in?

9. Personal taxation is usually structured in the following way:–

> no taxation on the first m_0 units of income;
> taxation at t_1% on the next m_1 units;
> taxation at t_2% on the next m_2 units;
> taxation at t_3% on anything above.

For some reason, this is termed *progressive* taxation. Write a generalised program to determine net income after tax deductions. Write out the gross income, the deductions and the net income. You will have to make some realistic estimates of the tax thresholds m_i and the taxation levels t_i. You could use this sort of model to find out how sensitive revenue from taxation was in relation to cosmetic changes in thresholds and tax rates.

11. The specific heat capacity of water is 2009 J kg^{-1} K^{-1}; the specific latent heat of fusion (ice/water) is 335 kJ kg^{-1}, and the specific latent heat of vaporization (water/steam) is 2500 kJ kg^{-1}. Assume that the specific heat capacity of ice and steam are identical to that of water. Write a program which will read in two temperatures, and will calculate the energy required to raise (or lower) ice, water or steam at the first temperature, to ice, water or steam at the second. Take the freezing point of water as 273 K, and its boiling point as 373 K. For those happier with Celsius, 0° C is 273 K, while 100° c is 373 K. One calorie is 4.1868 J, and for the truly atavistic, 1 BTU is 1055 J (approximately).

15.5 Bibliography

Dahl O. J., Dijkstra E. W., Hoare C. A. R., *Structured Programming*, Academic Press, 1972.

- This is the original text, and a must. The quote at the start of the chapter by Dijkstra summarises beautifully our limitations when programming and the discipline we must have to successfully master programming.

Knuth D. E., *Structured Programming with GOTO Statements,* in Current Trends in Programming Methodology, Volume 1, Prentice Hall.

- The chapter by Knuth provides a very succinct coverage of the arguments for the adoption of structured programming, and dispells many of the myths concerning the use of the GOTO statement. Highly recommended.

16
Characters

These metaphysics of magicians,
And necromantic books are heavenly;
Lines, circles, letters and characters;

Christopher Marlowe, *The Tragical History of Doctor Faustus*

Aims

The aims of this chapter are:–

- to extend the ideas about characters introduced in earlier chapters;
- to demonstrate that this enables us to solve a whole new range of problems in a satisfactory way;

16 Character

For each type in a programming language there are following concepts:-

- values are drawn from a finite domain
- there are a restricted number of operations defined for each type

For the numeric types we have already met, integers and reals

- the values are either drawn from the domain of integer numbers or the domain of real numbers
- the valid operations are addition, subtraction, multiplication, division and exponentiation.

For the character data type the basic unit is an individual character — any character which is available on your keyboard normally. To ensure portability we should restrict ourselves to the Fortran character set, that is:-

- the alphabetic characters A through Z
- the digits or numeric characters 0 through 9
- the underscore character _

and these may be used in variable names, and the special characters:-

	blank	)	right brackets or parenthesis	"	quotation mark
=	equal			%	percent
+	plus	,	comma	&	ampersand
-	minus	.	decimal point or period	;	semi-colon
*	asterisk			<	less than
/	slash or oblique	'	apostrophe	>	greater than
(	left brackets or parenthesis	:	colon	?	question mark
		!	exclamation mark	$	currency symbol

This provides us with 58 printing characters and omits many commonly used characters, e.g. lower case letters. However if one does work with this set then one can ensure that programs are portable.

As the most common current internal representation for the character data type uses 8 bits this should provide access to 256 (2^8) characters. However there is little agreement over the encoding of these 256 possible characters, and the best you can normally assume is access to the ASCII character set, which is given in appendix B. One of the problems looks at determining what characters one has available.

The only operations defined are concatenation (joining character strings together) and comparison.

We will look into the area of character sets in more depth later in this chapter.

We may declare our character variables:–

```
CHARACTER :: A, String, Line
```

Notice that there is no default typing of the character variable (unlike integer and real data types), and we can use any convenient name within the normal Fortran conventions. In the declaration above, each character variable would have been permitted to store one character. This is limiting, and, to allow character strings which are several units long, we have to add one item of information:–

```
CHARACTER (10) :: A
CHARACTER (16) :: String
CHARACTER (80) :: Line
```

This indicates that A holds 10 characters, STRING holds 16, and LINE holds 80. If all the character variables in a single declaration contain the same number of characters, we may abbreviate the declaration to

```
CHARACTER(80) :: LIST, STRING, LINE
```

But we cannot mix both forms in the one declaration. We can now assign data to these variables, as follows:–

```
A='FIRST ONE '
STRING='A LONGER ONE         '
LINE='THE QUICK BROWN FOX JUMPS OVER THE LAZY DOG'
```

The delimiter apostrophe (') or quotation mark(") is needed to indicate that this is a character string (otherwise the assignments would have looked like invalid variable names).

16.1 Character Input

In an earlier chapter we saw how we could use the READ * and PRINT * statements to do both numeric and character input and output or i/o. When we use this form of the statement we have to include any characters we type within delimiters (either the apostrophe ' or quotation mark "). This is a little restricting and there is a slightly more complex form of the READ statement that allows one to just type the string on its own. The following two programs illustrate the differences.

```
PROGRAM Chapter16_01
!
! Simple character i/o
!
```

```
CHARACTER (80) :: Line
   READ *, Line
   PRINT *, Line
END PROGRAM Chapter16_01
```

This form requires enclosing the string with delimiters. Consider the next form:–

```
PROGRAM Chapter16_02
!
! Simple character i/o
!
CHARACTER (80) :: Line
   READ '(A)' , Line
   PRINT *,Line
END PROGRAM Chapter16_02
```

With this form one can just type the string in and input terminates with the carriage return key. The additional syntax involves '(A)' where '(A)' is a character edit descriptor. The simple examples we have used so far have used implied format specifiers and edit descriptors. For each data type we have one or more edit descriptors to chose from. For the character data type only one edit descriptor is available and that is the A edit descriptor.

16.2 Character Operators

The first manipulator is a new operator — the concatenation operator //. With this operator we can join two character variables to form a third, as in

```
CHARACTER (5) :: FIRST, SECOND
CHARACTER (10) :: THIRD
FIRST='THREE'
SECOND='BLIND'
...
THIRD=FIRST//SECOND
.
THIRD=FIRST//'MICE'
```

Where there is a discrepancy between the created length of the concatenated string and the declared lengths of the character strings, truncation will occur. For example

```
THIRD=FIRST//' BLIND MICE'
```

will only append the first five characters of the string ' BLIND MICE' – that is ' BLIN', and THIRD will therefore contain 'THREE BLIN'.

What would happen if we assigned a character variable of length 'n' a string which was shorter than n? e.g.

```
CHARACTER (4) :: C2
C2='AB'
```

The remaining two characters are considered to be blank, that is, it is equivalent to saying

```
C2='AB  '
```

However, while the strings 'AB' and 'AB ' are equivalent, 'AB' and ' AB' are not. In the jargon, the character strings are always *left justified*, and the *unset* characters are trailing blanks.

If we concatenate strings which have 'trailing blanks', the blanks, or spaces, are considered to be legitimate characters, and the concatenation begins after the end of the first string. Thus

```
CHARACTER (4) :: C2,C3
CHARACTER (8) :: JJ
C2='A'
C3='MAN'
JJ=C2//C3
PRINT*, 'THE CONCATENATION OF ',C2,' AND ',C3,' IS'
PRINT*,JJ
```

would appear as

THE CONCATENATION OF A   AND MAN GIVES

A   MAN

at the terminal.

Sometimes we need to be able to extract parts of character variables — sub-strings. The actual notation for doing this is a little strange at first, but it is very powerful. To extract a sub-string we must reference two items;

- (i) the position in the string at which the sub-string begins,

and

- (ii) the position at which it ends.

e.g.

```
STRING='SHARE AND ENJOY'
```

16.3 Character Sub-Strings

We may extract parts of this string

```
BIT=STRING(3:5)
```

would place the characters 'ARE' into the variable BIT. This may be manipulated further

```
BIT1=STRING(2:4)//STRING(9:9)
BIT2=STRING(5:5)//STRING(3:3)//STRING(1:1)//STRING(15:15)
```

Note that to extract a *single* character we reference its beginning position and its end (i.e. repeat the same position), so that

```
STRING(3:3)
```

gives the single character 'A'. The sub-string reference can cut out either one of the two numerical arguments. If the first is omitted, the characters up to and including the reference are selected, so that

```
SUB=STRING(:5)
```

would result in SUB containing the characters 'SHARE'. When the second argument is omitted, the characters from the reference are selected, so that

```
SUB=STRING(11:)
```

would place the characters 'ENJOY' in the variable SUB. In these examples it would also be necessary to declare STRING, SUB, BIT, BIT1 and BIT2 as CHARACTER type, of some appropriate length.

Character variables may also form arrays.

```
CHARACTER (10) , DIMENSION(20) :: A
```

sets up a character array of twenty elements, where each element contains ten characters. In order to extract sub-strings from these array elements, we need to know where the array reference and the sub-string reference are placed. The array reference comes first, so that

```
DO I=1,20
   FIRST=A(I)(1:1)
ENDO
```

places the first character of each element of the array into the variable FIRST. The syntax is therefore 'position in array, followed by position within string'.

Any argument can be replaced by a variable:

```
STRING(I:J)
```

This offers interesting possibilities, since we can, for example, strip out blanks from a string

```
      CHARACTER(80) ::    STRING, STRIP
      INTEGER :: IPOS,I,LEN
      IPOS=0
      DO I=1,LEN
         IF(STRING(I:I) /= ' ') THEN
            IPOS=IPOS+1
            STRIP(IPOS:IPOS)=STRING(I:I)
         ENDIF
      END DO
      PRINT*,STRING
      PRINT*,STRIP
```

16.4 Character functions

There are special functions available for use with character variables: INDEX will give the starting position of a string within another string. If, for example we were looking for all occurrences of the string 'Geology' in a file, we could construct something like

```
PROGRAM Chapter16_03
IMPLICIT NONE
CHARACTER (80) :: Line
INTEGER :: I
   DO
      READ '(A)', Line
      I=INDEX(Line,'Geology')
      IF (I /= 0) THEN
         PRINT *, ' String Geology found at position ', I
         PRINT *, ' in line ', Line
         EXIT
      ENDIF
   ENDDO
END PROGRAM Chapter16_03
```

There are two things to note with the INDEX function. Firstly, it will only report the first occurrence of the string in the line; any later occurrences in any particular line will go unnoticed, unless you account for this in some way. Secondly, if the string does not occur, the result is zero, and hence the form of the logical expression in the IF THEN ENDIF statement.

LEN provides the length of a character string. This function is not immediately useful, since you really ought to know how many characters there are in the string. However, as later examples will show, there are some cases where it can be useful. Remember that trailing blanks do count as part of the character string, and contribute to the length.

The next group of functions need to be considered together. They revolve around the concept of a collating sequence. In other words, each character used in Fortran is ordered as a list, and given a corresponding *weight*. No two weights are equal. Although Fortran has only 58 defined characters, the machine you use will generally have more; 95 printing characters a typical minimum number. On this type of machine the weights would vary from zero to 94. There is a defined collating sequence, the ASCII sequence, which is likely to be the default. The parts of the collating sequence which are of most interest are fairly standard throughout all collating sequences.

In general, we are interested in the numerals (0–9), the alphabetics (A–Z) and a few odds and ends like the arithmetic operators (+ – / *), some punctuation (. and ,) and perhaps the prime ('). As you might expect, 0–9 carry successively higher weights (though not the weights 0 to 9), as do A to Z. The other odds and ends are a little more problematic, but we can find out the weights through the function ICHAR. This function takes a single character as argument, and returns an integer value. The ASCII weights for the alphanumerics are as follows:–

```
0-9    48-57
A-Z    65-90
```

One of the exercises is to determine the weights for other characters. The reverse of this procedure is to determine the character from its weighting, which can be achieved through the function CHAR. CHAR takes an integer argument and returns a single character. Using the ASCII collating sequence, the alphabet would be generated from

```
DO   I=65,90
   PRINT*,CHAR(I)
ENDDO
```

This idea of a weighting may then be used in four other functions:–

Function	Action
LLE	lexically less than or equal to
LGE	lexically greater than or equal to
LGT	lexically greater than
LLT	lexically less than

In the sequence we have seen before, A is lexically less than B; i.e. its weight is less. Clearly, we can use ICHAR and get the same result. For example

```
IF(LGT('A','B'))  THEN
```

is equivalent to

```
IF(ICHAR('A') > ICHAR('B')) THEN
```

but these functions can take character string arguments of any length. They are not restricted to single characters.

These functions provide very powerful tools for the manipulation of characters, and open up wide areas of non-numerical computing through Fortran. Lots of text formatting and word processing applications may now be tackled (conveniently ignoring the fact that lower case characters may not be available).

There are many problems that require the use of character variables. These range from the ability to provide simple titles on reports, or graphical output, to the provision of a natural language interface to one of your programs, i.e. the provision of an English-like command language. *Software Tools*, Kernighan and Plauger contains many interesting uses of characters in Fortran.

16.5 Summary

- Characters represent a different data type to any other in Fortran, and as a consequence there is a restricted range of operations which may be carried out on them.

- A character variable has a length which must be assigned in a CHARACTER declaration statement.

- Character strings are delimited by apostrophes (') or quotation marks ("). Within a character string, the blank is a significant character.

- Character strings may be joined together (concatenated) with the // operator.

- Sub-strings, occurring within character strings, may be also be manipulated. There are a number of functions especially for use with characters — ACHAR, ADJUSTL, ADJUSTR, CHAR, IACHAR, INDEX, LEN, LEN_TRIM, LLE, LGE, LGT, LLT, REPEAT, SCAN, TRIM AND VERIFY.

16.6 Problems

1. Suggest some circumstances where PRIME='''' might be useful. What other alternative is there and why do you think we use that instead?

2. Write a program to write out the weights for the Fortran character set. Modify this program to print out the weights of the complete implementation defined character set for your version of Fortran 90. Is it ASCII? If not how does it differ?

3. Use the INDEX function in order to find the location of all the strings 'IS' in the following data;

IF A PROGRAMMER IS FOUND TO BE INDISPENSABLE, THE BEST THING TO DO IS TO GET RID OF HIM AS QUICKLY AS POSSIBLE.

4. Find the 'middle' character in the following strings. Do you include blanks as characters? What about punctuation?

PRACTICE IS THE BEST OF ALL INSTRUCTORS. EXPERIENCE IS A DEAR TEACHER, BUT FOOLS WILL LEARN AT NO OTHER.

5. In English, the order of occurrence of the letters, from most frequent to least is:–

E, T, A, O, N, R, I, S, H, D, L, F, C, M, U, G, Y, P, W, B, V, K, X, J, Q, Z.

Use this information to examine the two files given in appendix D (one is a translation of the other) to see if this is true for these two extracts of text. The second text is in medieval Latin (c. 1320). Note that a fair amount of compression has been achieved by expressing the passage in Latin rather than modern English. Does this provide a possible model for information compression?

6. A very common cypher is the substitution cypher, where, for example, every letter A is replaced by (say) an M, every B is replaced by (say) a Y, and so on. These encyphered messages can be broken by reference to the frequency of occurrence of the letters (given in the previous question). Since we know that (in English) E is the most commonly occurring letter, we can assume that the most commonly occurring letter in the encyphered message represents an E; we then repeat the process for the next most common and so on. Of course, these correspondences may not be exact, since the message may not be long enough to develop the frequencies fully. However, it may provide sufficient information to break the cypher. The file given in Appendix E contains an encoded message. Break it. Clue — *Pg Fybdujuvef jo Tdjfodf,* Jorge Luis Borges.

17
Complex

Make it as simple as possible, but no simpler.

Albert Einstein

'Can you do addition?' the White Queen asked. 'What's one and one and one and one and one and one and one and one and one?' 'I don't know' said Alice. 'I lost count.' 'She can't do addition,' the Red Queen interrupted.

Lewis Carroll, *Through the Looking Glass and What Alice Found There*.

Aims

The aims of this chapter are:–

- to introduce the last predefined numeric data type in Fortran;
- to illustrate with examples how to use this type.

17 Complex

This variable type reflects an extension of the real data type available in Fortran – the COMPLEX data type, where we can store and manipulate complex variables. Problems that require this data type are restricted to certain branches of mathematics, physics and engineering. Complex numbers are defined as having a *real* and *imaginary* part; i.e.:–

$$a = x + iy$$

where i is the square root of –1.

They are not supported in many programming languages as a base type. This therefore makes Fortran the language of first choice for many people.

To use this variable type we have to write the number as two parts, the real and imaginary elements of the number, for example:–

```
COMPLEX :: U
   U=(1.0,2.0)
```

represents the complex number 1+i2. Note that the complex number is enclosed in brackets. We can do arithmetic on variables like this, and most of the intrinsic functions like LOG, SIN, COS etc. accept complex data type as argument.

All the usual rules about mixing different variable types, like reals and integers, also apply to complex. Complex numbers are read in and written out in a similar way to real numbers, but with the provision that, for each single complex value, two format descriptors must be given. You may use either E or F formats (or indeed, mix them), as long as there are enough of them. Although you use brackets around the pairs of numbers in a program, these must not appear in any input, nor will they appear on the output.

Fortran has a number of functions which help to clarify the intent of *mixed mode* expressions. The functions REAL, CMPLX and INT can be used to 'force' any variable to real, complex or integer type.

There are a number of intrinsic functions to enable complex calculations to be performed. For example if we have:–

```
COMPLEX:: Z, Z1,Z2,Z3,ZBAR
REAL :: X,Y,X1,Y1,X2,Y2,X3,Y3,ZMOD
   Z1 = CMPLX (1.0, 2.0)         ! 1 + i 2
   Z2 = CMPLX(X2, Y2)            ! X2 + i Y2
   Z3 = CMPLX (X3, Y3)           ! X3 + i Y3
   Z  = Z1*Z2 / Z3
   X  = REAL(Z)                  ! real part of Z
   Y  = AIMAG (Z)                ! imaginary part of Z
```

```
            ZMOD = ABS(Z)                ! modulus of Z
            ZBAR = CONJG(Z)              ! complex conjugate of Z
```

17.1 Example

The second order differential equation:–

$$\frac{d^2 y}{dt^2} + 2\frac{dy}{dt} + y = x(t)$$

could describe the behaviour of an electrical system, where x(t) is the input voltage and y(t) is the output voltage and dy/dt is the current. The complex ratio

$$\frac{y(w)}{x(w)} = 1 / (-w^2 + 2jw + 1)$$

is called the frequency response of the system because it describes the relationship between input and output for sinusoidal excitation at a frequency of w and where j is $\sqrt{-1}$. The following program segment reads in a value of w and evaluates the frequency response for this value of w together with its polar form (magnitude and phase).

```
PROGRAM Complex_numbers
IMPLICIT NONE
!
! Program to calculate frequency response of a system
! for a given Omega
! and its polar form (magnitude and phase).
!
REAL    :: Omega,Real_part,Imag_part,Magnitude, Phase
COMPLEX:: Frequency_response
!
! Input frequency Omega
!
   PRINT *, 'Input frequency'
   READ *,Omega
!
   Frequency_response=1.0/CMPLX( - Omega * Omega + 1.0 , 2.0*Omega)
   Real_part = REAL(Frequency_response)
   Imag_part = AIMAG(Frequency_response)
!
! Calculate polar coordinates (magnitude and phase)
!
   Magnitude = ABS(Frequency_response)
   Phase = ATAN2 (Imag_part, Real_part)
```

```
    !
    PRINT *, ' At frequency ',Omega
    PRINT *, 'Response = ', Real_part,' + I ',Imag_part
    PRINT *,'in Polar form'
    PRINT *,' Magnitude = ', Magnitude
    PRINT *,' Phase = ', Phase
END PROGRAM Complex_numbers
```

17.2 Complex and Kind Type

The standard requires that there be a minimum of two kind types for real numbers and this is also true of the complex data type. Chapter 8 must be consulted for a full coverage of real kind types. We would therefore use something like the following to select a complex kind type other than the default:–

```
INTEGER , PARAMETER :: Long_Complex=SELECTED_REAL_KIND(15,307)
COMPLEX (Long_Complex) :: Z
```

Chapter 26 contains a good example of how to use modules to define and use precision throughout a program and sub-program units.

17.3 Summary

COMPLEX is used to store and manipulate complex numbers those with a real and imaginary part.

There are standard functions which allow conversion between the numerical data types — CMPLX, REAL and INT.

17.4 Problems

1. The program used in chapter 15 which calculated the roots of a quadratic had to abandon the calculation if the roots were complex. You should now be able to remedy this, remembering that it is necessary to declare any complex variables. Instead of raising the expression to the power 0.5 in order to square root it, use the function SQRT. If you manage this to your satisfaction, try your skills on the roots of a cubic (see the problems in chapter 15).

18
Logical

A messenger yes/no semaphore
her black/white keys in/out whirl of morse
hoopooe signals salvation deviously

Nathaniel Tarn, *The Laurel Tree*

Aims

The aims of this chapter are:–

- to examine the last predefined type available in Fortran: logical
- to introduce the concepts necessary to use logical expressions effectively; namely
    - logical variables
    - logical operators
    - the hierarchy of operations
    - truth tables

18 Logical

Often we have situations where we need ON/OFF, TRUE/FALSE or YES/NO switches, and in such circumstances we can use LOGICAL type variables: e.g.

```
LOGICAL :: FLAG
```

Logicals may take only two possible values, as shown following

```
FLAG=.TRUE.
```

or

```
FLAG=.FALSE.
```

Note the full stops, which are essential. With a little thought you can see why they are needed. You will already have met some of the ideas associated with logical variables from IF statements.

```
IF (A == B) THEN
    .
ELSE
    .
ENDIF
```

The logical expression (A == B) returns a value *true* or *false,* which then determines the route to be followed; if the quantity is true, then we execute the next statement, else we take the other route.

Similarly, the following example is also legitimate:–

```
LOGICAL :: ANSWER
    ANSWER=.TRUE.
    ...
    IF (ANSWER) THEN
        ...
    ELSE
        ...
    ENDIF
```

Again the expression IF (ANSWER) is evaluated; here the variable ANSWER has been set to .TRUE., and therefore the statements following the THEN are executed. Clearly, conventional arithmetic is inappropriate with logicals. What does 2 times true mean? (very true?). There are a number of special operators for logicals:

.NOT. which negates a logical value (i.e. changes *true* to *false* or vice versa)

.AND. logical union

.OR. logical intersection

To illustrate the use of these operators, consider the following program extract:

```
LOGICAL  ::  A,B,C
    A=.TRUE.
    B=.NOT.A
!                       (B now has the value 'false')
    C=A.OR.B
!                       (C has the value 'true')
    C=A.AND.B
!                       (C now has the value 'false')
```

To gauge the effect of these operators on logicals, we can consult a truth table:–

X1	X2	.NOT.X1	X1.AND.X2	X1.OR.X2
true	true	false	true	true
true	false	false	false	true
false	true	true	false	true
false	false	true	false	false

As with arithmetic operators, there is an order of precedence associated with the logical operators.

.AND. is carried out before

.OR. and .NOT.

In dealing with logicals, the operations are carried out within a given level, from left to right. Any expressions in brackets would be dealt with first. The logical operators are a lower order of precedence to the arithmetic operators, i.e. they are carried out later. A more complete operator hierarchy is therefore:

expressions within brackets
exponentiation
multiplication/division
addition/subtraction
relational logical (= =, >, <, >=, <= /=)
.AND.
.OR. and .NOT.

Although you can build up complicated expressions with mixtures of operators, these are often difficult to comprehend, and it is generally more straightforward to break 'big' expressions down into smaller ones, whose purpose is more readily appreciated.

Logical

Historically, logicals have not been in evidence extensively in Fortran programs, although clearly there are occasions on which they are of considerable use. Their use often aids considerably in making programs more modular and comprehensible. They can be used to make a complex section of code involving several choices much more transparent by the use of one logical function, with an appropriate name. Logicals may be used to control output, e.g.

```
LOGICAL :: DEBUG
   ...
   DEBUG=.TRUE.
   ...
   IF(DEBUG)THEN
      ...
      PRINT *,'LOTS OF PRINTOUT'
      ...
   ENDIF
```

ensures that, while de-bugging a program you have more output. Then, when the program is *correct*, run with DEBUG=.FALSE.

Note that Fortran does try to protect you while you use logical variables. You cannot do this:

```
LOGICAL :: UP, DOWN
   UP=DOWN+.FALSE.
```

or

```
LOGICAL :: A2
   REAL DIMENSION(10):: OMEGA
   .
   A2=OMEGA(3)
```

The compiler will note that this is an error, and will not permit you to run the program. This is an example of *strong typing*, since only a limited number of predetermined operations are permitted. The real, integer and complex variable types are much more weakly typed (which helps to lead to the confusion inherent in mixing variable types in arithmetic assignments).

18.1 I/O

Since logicals may take only the values .TRUE. and .FALSE., the possibilities in reading and writing logical values are clearly limited. The L format allows logicals to be input and output. On input, if the first non-blank characters are either T or .T, the logical value .TRUE. is stored in the corresponding list item; if the first non-blank characters are F or .F, then .FALSE. is stored. (Note therefore that reading, say, TED and FAHR in an L4 format would be acceptable.) If the first non-blank character is not F, T, .F or .T, then an error message will be generated. On output, the value T or F is written out, right justified, with blanks (if appropriate). Thus,

```
LOGICAL :: FLAG
   FLAG=.TRUE.
   PRINT 100, FLAG, .NOT.FLAG
   100 FORMAT(2L3)
```

would produce

```
  T  F
```

at the terminal.

Assigning a logical variable to anything other than a .TRUE. or .FALSE. value in your program will result in errors. The 'shorthand' forms of .T, .F, F and T are not acceptable in the program.

18.2 Summary

Another type of data — logical — is also recognised. A LOGICAL variable may take one of two values — *true* or *false*.

- There are special operators for manipulating logicals .NOT., .AND. and .OR..
- Logical operators have a lower order of precedence than any others.

18.3 Problems

1. Why are the full stops needed in a statement like A = .TRUE.?

2. Generate a truth table like the one given in this chapter.

3. Write a program which will read in numerical data from the terminal, but will *flag* any data which is negative, and will also turn these negative values into positive ones.

19
User Defined Types

Russell's theory of types leads to certain complexities in the foundations of mathematics,... Its interesting features for our purposes are that types are used to prevent certain erroneous expressions from being used in logical and mathematical formulae; and that a check against violation of type constraints can be made purely by scanning the text, without any knowledge of the value which a particular symbol might happen to have.

C. A. R. Hoare, *Structured Programming*.

Aims

The aim of this chapter is to introduce the concepts and ideas involved in using the facilities offered in Fortran 90 for the construction and use of user defined types:–

- the way in which we define our own types;
- the way in which we declare variables to be of a user defined type;
- the way in which we manipulate variables of our own types;
- the way in which we can nest types within types.

The examples are simple and are designed to highlight the syntax. More complex and realistic examples of the use of user defined data types are to be found in later chapters.

19 User Defined Types

In the coverage so far we have used the intrinsic types provided by Fortran. The only data structuring technique available has been to construct arrays of these intrinsic types. Whilst this enables us to solve a reasonable variety of problems, it is inadequate for many purposes. In this chapter we look at the facilities offered by Fortran for the construction of our own types, and how we manipulate data of these new, user defined types.

With the ability to define our own types we can now construct aggregate data types that have components of a variety of base types. These are often given the name records in books on data structures. In mathematics the term cartesian product is often used, and this is the terminology adopted by Hoare. We will stick to the term records, as it is the one that is most commonly used in computing and texts on programming.

There are two stages in the process of creating and using our own data types, we must first define the type, and secondly create variables of this type.

19.1 Example 1 – Dates

```
PROGRAM Record_01
IMPLICIT NONE
TYPE Date
   INTEGER :: Day
   INTEGER :: Month
   INTEGER :: Year
END TYPE Date
TYPE (Date) :: D
   PRINT *,' Type in the date, day, month, year'
   READ *,D%Day, D%Month, Day%Year
   PRINT *,D%Day, D%Month, Day%Year
END PROGRAM Record_01
```

This complete program illustrates both the definition and use of the type.

19.2 Type Definition

The type *Date* is defined to have three component parts, comprising a *day*, a *month* and a *year*, all of integer type. The syntax of a type construction comprises:–

TYPE Typename
  Data Type :: Component_name
  etc
END TYPE Typename

Reference can then be made to this new type by the use of a single word, *Date*, and we have a very powerful example of the use of abstraction.

User Defined Types

19.3 Variable Definition

This is done by

`TYPE (Typename) :: Variablename`

and we then define a variable *D* to be of this new type. The next thing we do is have a READ * statement that prompts the user to type in three integer values, and the data is then echoed straight back to the user. We use the notation `Variablename%Component_Name` to refer to each component of the new data type.

19.4 Example 2 – Address lists

Here is the complete program.

```
PROGRAM Record_02
IMPLICIT NONE
TYPE Address
   CHARACTER (LEN=40) :: Name
   CHARACTER (LEN=60) :: Street
   CHARACTER (LEN=60) :: District
   CHARACTER (LEN=60) :: City
   CHARACTER (LEN=8)  :: Post_Code
END TYPE Address
INTEGER , PARAMETER :: N_of_Address=78
TYPE (Address) , DIMENSION(N_of_Address):: Addr
INTEGER :: I
   OPEN(UNIT=1,FILE="ADDRESS.DAT")
   DO I=1,N_of_Address
      READ(UNIT=1,FMT='(A40)') Addr(I)%Name
      READ(UNIT=1,FMT='(A60)') Addr(I)%Street
      READ(UNIT=1,FMT='(A60)') Addr(I)%District
      READ(UNIT=1,FMT='(A60)') Addr(I)%City
      READ(UNIT=1,FMT='(A8)') Addr(I)%Post_Code
   END DO
   DO I=1,N_of_Address
      PRINT *,Addr(I)%Name
      PRINT *,Addr(I)%Street
      PRINT *,Addr(I)%District
      PRINT *,Addr(I)%City
      PRINT *,Addr(I)%Post_Code
   END DO
END PROGRAM Record_02
```

In this example we define a type Address which has components that one would expect for a person's address. We then define an array Addr of this type. Thus we are now creating arrays of our own user defined types. We index into the array in the way we would expect from our experience with integer, real and character arrays. The complete example is rather trivial in a sense in that the program merely reads from one file and prints the file out to the screen. However it highlights many of the important ideas of the definition and use of user defined types.

19.5 Example 3: Nested User Defined Types

The following example builds on the two data types already introduced. Here we construct nested user defined data types based on these types and construct a new data type containing them both, plus additional information.

```
PROGRAM C19_03
IMPLICIT NONE
TYPE Address
   CHARACTER (LEN=60)    :: Street
   CHARACTER (LEN=60)    :: District
   CHARACTER (LEN=60)    :: City
   CHARACTER (LEN=8 )    :: Post_Code
END TYPE Address
TYPE Date_Of_Birth
   INTEGER :: Day
   INTEGER :: Month
   INTEGER :: Year
END TYPE Date_Of_Birth
TYPE Personal
   CHARACTER (LEN=20)      :: First_Name
   CHARACTER (LEN=20)      :: Other_Names
   CHARACTER (LEN=40)      :: Surname
   TYPE (Date_Of_Birth)    :: DOB
   CHARACTER (LEN=1)       :: Sex
   TYPE (Address)          :: Addr
END TYPE Personal
INTEGER , PARAMETER :: N_People=2
TYPE (Personal) , DIMENSION(N_People) :: P
INTEGER :: I
   OPEN(UNIT=1,FILE='PERSON.DAT')
   DO I=1,N_People
      READ(1,FMT=10) P(I)%First_Name,&
                     P(I)%Other_Names,&
                     P(I)%Surname,&
                     P(I)%DOB%Day,&
```

User Defined Types

```
                        P(I)%DOB%Month,&
                        P(I)%DOB%Year,&
                        P(I)%Sex,&
                        P(I)%Addr%Street,&
                        P(I)%Addr%District,&
                        P(I)%Addr%City,&
         P(I)%Addr%Post_Code
   10 FORMAT(  A20,/,&
               A20,/,&
               A40,/,&
               I2,1X,I2,1X,I4,/,&
               A1,/,&
               A60,/,&
               A60,/,&
               A60,/,&
               A8)
      END DO
      DO I=1,N_People
         WRITE(*,FMT=20)   P(I)%First_Name,&
                        P(I)%Other_Names,&
                        P(I)%Surname,&
                        P(I)%DOB%Day,&
                        P(I)%DOB%Month,&
                        P(I)%DOB%Year,&
                        P(I)%Sex,&
                        P(I)%Addr%Street,&
                        P(I)%Addr%District,&
                        P(I)%Addr%City,&
                        P(I)%Addr%Post_Code
   20 FORMAT(       A20,A20,A40,/,&
                    I2,1X,I2,1X,I4,/,&
                    A1,/,&
                    A60,/,&
                    A60,/,&
                    A60,/,&
                    A8)
      END DO
END PROGRAM C19_03
```

Here we have a date of birth data type (Date_Of_Birth) based on the Date data type from the first example, plus a slightly modified address data type, incorporated into a new data type comprising personal details. Note the way in which we reference the component parts of this new, aggregate data type.

19.6 Problems

1. Modify the last example to include a more elegant printed name. The current example will pad with blanks the first name, other names and surname and span 80 characters on one line, which looks rather ugly.

Add a new variable name which will comprise all three sub-components and write out this new variable, instead of the three sub-components.

19.7 Bibliography

Dahl O. J., Dijkstra E. W., Hoare C. A. R., *Structured Programming*, Academic Press, 1972.

- This is one of the earliest and best introductions to data structures and structured programming. The whole book hangs together very well, and the section on data structures is a must for serious programmers.

Fortran 90 is a relatively young language, and there are a lack of books that look at data structuring using Fortran 90 as the vehicle. This is not the problem that it might appear at first sight, as there a wealth of books that are in the Algol, Pascal, Modula 2 family of languages, that are cheap and widely available. Some of these are given below.

Kruse R.L., *Data Structures and Program Design*, Prentice Hall

- Quite a gentle introduction to the use of recursion and its role in problem solving. Good choice of case studies with explanations of solutions. Pascal is used.

Sedgewick R., *Algorithms*, Addison Wesley.

- Good source of algorithms. Well written. The GCD algorithm was taken from this source.

Wirth N., *Algorithms + Data Structures = Programs*, Prentice Hall.

Wirth N., *Algorithms + Data Structures*, Prentice Hall.

- The first is in Pascal, and the second in Modula 2.

Wood D., *Paradigms and Programming in Pascal*, Computer Science Press.

- Contains a number of examples of the use of recursion in problem solving. Also provides a number of useful case studies in problem solving.

20
Dynamic Data Structures

The question naturally arises whether the analogy can be extended to a data structure corresponding to recursive procedures. A value of such a type would be permitted to contain more than one component that belongs to the same type as itself; in the same way that a recursive procedure can call itself recursively from more than one place in its own body.

C. A. R. Hoare, *Structured Programming*.

Aims

The primary aims of the chapter are to introduce the facilities offered by Fortran 90 for the construction and use of dynamic data types. These additional features enable us to solve a much wider range of problems using simple and straightforward algorithms. Examples include the use of pointers in user defined data types, in particular:–

- singly linked lists;
- trees;
- using singly linked lists to store sparse vectors;

The secondary aims are to provide a summary of what can be achieved with the correct choice of data structures and algorithms, and the bibliography contains details sources of data structures and algorithms.

20 Dynamic Data Structures

All of the data types introduced so far, with the exception of the allocatable array, have been static. Even with the allocatable array a size has to be set at some stage during program execution. The facilities provided in Fortran by the concept of a pointer combined with those offered by a user defined record enable us to address a completely new problem area, previously extremely difficult to solve in Fortran.

There are many problems where one genuinely does not know what requirements there are on the size of a data structure. Anyone who has used a word processing package or editor would not like to have to provide information at the start about the size of their document or file.

20.1 Example 1: Simple Pointer Concepts

With the introduction of pointers as a data type into Fortran 90 we also have the introduction of a new assignment statement, the pointer assignment statement. The following example highlights clearly the main concepts involved with pointer use.

```
PROGRAM C20_01
   INTEGER , POINTER :: A,B
   INTEGER , TARGET :: C
   INTEGER :: D
   C = 1
   A => C
   C = 2
   B => C
   D = A + B
   PRINT *,A,B,C,D
END PROGRAM C20_01
```

The first declaration defines A and B to be variables, with the POINTER attribute. This means we can use A and B to refer or point to integer values.

The second declaration defines C to be an integer, with the TARGET attribute, i.e. we can use pointers to refer or point to the value of the variable C.

The last declaration defines D to be an ordinary integer variable.

Let us now look at the various executable statements in the program, one at a time:−

C = 1	This is an example of the normal assignment statement that we are already familiar with. We use the variable name C in our program and whenever we use that name we get the *value* of the variable C.
A => C	This is an example of a pointer assignment statement. This means that both A and C now refer to the same value, in this case 1.
C = 2	Conventional assignment statement, and C now has the value 2.

Dynamic Data Structures

B = > C   Second example of pointer assignment. B now points to the value that C has, in this case 2.

D = A + B   Simple arithmetic assignment statement. The value that A points to is added to the value that B points to and the result is assigned to D.

The last statement prints out the values of A, B, C and D.

The output is:–

2 2 2 4

and this complete highlights the essentials of pointer usage, and also the difference between normal assignment and pointer assignment.

Let us look at a simple example that will read an arbitrary amount of text from the keyboard, and then echo it back to the user.

20.2 Example 2: Singly linked list

Conceptually a singly linked lists consists of a sequence of boxes with two compartments. The first compartment holds a data item and the second compartment contains directions to the next box. Graphically:–

We can construct a data structure in Fortran to work with a singly linked list by combining the concept of a record from the previous chapter with the new concept of a pointer. A complete program to do this is given below.

```
PROGRAM C20_02
TYPE Link
   CHARACTER :: C
   TYPE (Link) , POINTER    :: Next
END TYPE Link
TYPE (Link) , POINTER :: Root , Current
INTEGER :: IO_Stat_Number=0
   ALLOCATE(Root)
   READ (UNIT=*,FMT='(A)',ADVANCE='NO',IOSTAT=IO_Stat_Number)Root%C
   IF (IO_Stat_Number == -1) THEN
      NULLIFY(Root%Next)
   ELSE
```

```
         ALLOCATE(Root%Next)
      ENDIF
      Current => Root
      DO WHILE (ASSOCIATED(Current%Next))
         Current => Current%Next
         READ (UNIT=*,FMT='(A)',ADVANCE='NO', &
         IOSTAT=IO_Stat_Number) Current%C
         IF (IO_Stat_Number == -1) THEN
            NULLIFY(Current%Next)
         ELSE
            ALLOCATE(Current%Next)
         ENDIF
      END DO
      Current => Root
      DO WHILE (ASSOCIATED(Current%Next))
         PRINT * , Current%C
         Current => Current%Next
      END DO
END PROGRAM C20_02
```

The first thing of interest is the type definition for the singly linked list. We have:–

```
TYPE Link
   CHARACTER :: C
   TYPE (Link) , POINTER :: Next
END TYPE Link
```

and we call the new type `Link`. It comprises two component parts, the first holds a character C, and the second holds a pointer called `Next` to allow us to refer to another instance of type `Link`. Remember we are interested in joining together several boxes or `Links`.

The next item of interest is the variable definition. Here we define two variables `Root` and `Current` to be pointers that point to items of type `Link`. In Fortran when we define a variable to be a pointer we also have to define what it is allowed to point to. This is a very useful restriction on pointers, and helps make using them more secure.

The first executable statement

```
ALLOCATE(Root)
```

requests that the variable `Root` be allocated memory. At this time the contents of both the character component and the pointer component are undefined.

The next statement reads a character from the keyboard. We are using a number of additional features of the READ statement, including:–

Dynamic Data Structures

```
ADVANCE='NO'
IOSTAT=IO_Stat_Number
```

and the two options combine to provide the ability to read an arbitrary amount of text from the user per line, and terminate only when end of file is encountered as the only input on a line, typically by typing CTRL Z. Please note that the numbers returned by the IOSTAT option are implementation specific. A small program would have to be written to test the values returned for each platform.

If an end of file is reached then `Root%Next` is nullified, and this is a convenient way of saying that it doesn't point to anything valid.

If the end of file is not detected then the next link in the chain is created.

The statement

```
Current => Root
```

means that both `Current` and `Root` point to the same physical memory location, and this holds a character data item and a pointer. We must do this as we have to know where the start of the list is. This is now our responsibility, not the compilers. Without this statement we would not be able to do anything with the list except fill it up – hardly very useful.

The WHILE loop is then repeated until end of file is reached. If the user had typed an end of file immediately then Current%Next would not be ASSOCIATED, and the WHILE loop would be skipped.

This loop allocates memory and moves down the chain of boxes one character at a time filling in the links between the boxes as we go.

We then have

```
Current => Root
```

and this now means that we are back at the start of the list, and in a position to traverse the list and print out each character in the list.

There is thus the concept with the pointer variable Current of it providing us with a window into memory where the complete linked list is held, and we look at one part of the list at a time.

It is recommended that this program be typed in, compiled and executed. It is surprisingly difficult to believe that this program will actually read in a completely arbitrary number of characters from the user. Seeing is believing.

20.3 Other Dynamic Data Structures

There are a wide range of dynamic data types and they include lists, queues, stacks and trees. Lists are a very commonly used data structure, especially in the doubly linked form where we can traverse in either direction. A queue is also a commonly used data structure and exists in a variety of forms, most noticeable *last in first out* (LIFO) and *first in first out* (FIFO). A stack is another quite widely used data structure. These three data structures are essentially one dimensional. A tree is a two dimensional data structure. It is one of the most widely used dynamic data structures and occurs in many algorithms. The Sedgewick book highlights the importance of the tree as a very versatile and powerful data structure.

20.4 Trees

The tree data structure occurs quite commonly in real life. One example is a family tree, shown below, another is in a knock out competition, e.g.

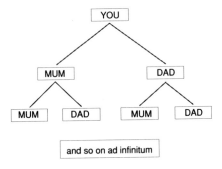

Family Tree

Trees have a surprisingly wide range of use in computer related applications, e.g. the compiler will construct a parse tree during the compilation process.

20.4.1 Example 3: Perfectly Balanced Tree

Let us now look at a more complex example that builds a perfectly balanced tree and prints the tree out. A loose definition of a perfectly balanced tree is a tree that for n nodes has minimum depth. More accurately a tree is perfectly balanced if for each node the number of nodes in its left and right sub-trees differ by at most 1.

```
MODULE Node_Type_Def
   TYPE Tree_Node
   INTEGER :: Number
   TYPE (Tree_Node) , POINTER :: Left,Right
   END TYPE Tree_Node
END MODULE Node_Type_Def

PROGRAM C20_03
! Construction of a perfectly balanced tree
```

Dynamic Data Structures

```fortran
      USE Node_Type_Def
      TYPE (Tree_Node) , POINTER :: Root
      INTEGER :: N_of_Items

      INTERFACE

         RECURSIVE FUNCTION Tree(N) RESULT(Answer)
            USE Node_Type_Def
            TYPE (Tree_Node) , POINTER :: Answer
            INTEGER :: N
         END FUNCTION Tree

         SUBROUTINE Print_Tree(Trees,H)
            USE Node_Type_Def
            TYPE (Tree_Node) , POINTER :: Trees
            INTEGER :: H
         END SUBROUTINE Print_Tree

      END INTERFACE

      PRINT *,' Enter number of items'
      READ *,N_Of_Items
      Root=>Tree(N_Of_Items)
      CALL Print_Tree(Root,0)

END PROGRAM C20_03

RECURSIVE FUNCTION Tree(N) RESULT (Answer)
USE Node_Type_Def
TYPE (Tree_Node) , POINTER :: Answer
TYPE (Tree_Node) , POINTER :: New_Node

INTEGER :: L,R,X
   IF (N == 0) THEN
      print *,' terminate tree'
      NULLIFY(Answer)
   ELSE
      L=N/2
      R=N-L-1
      PRINT *,L,R,N
      PRINT *,' Next item'
      READ *,X
```

```
         ALLOCATE(New_Node)
         New_Node%Number=X
         print *, ' left branch'
         New_Node%Left => Tree(L)
         print *, ' right branch'
         New_Node%Right => Tree(R)
         Answer => New_Node
      ENDIF
      PRINT *, ' Function tree ends'
END FUNCTION Tree

RECURSIVE SUBROUTINE Print_Tree(T,H)
USE Node_Type_Def
TYPE (Tree_Node) , POINTER :: T
INTEGER :: I
INTEGER :: H
      IF (ASSOCIATED(T)) THEN
         CALL Print_Tree(T%Left,H+1)
         DO I=1,H
            WRITE(UNIT=*,FMT=10,ADVANCE='NO')
            10 FORMAT('    ')
         ENDDO
         PRINT *,T%Number
         CALL Print_Tree(T%Right,H+1)
      ENDIF
END SUBROUTINE Print_Tree
```

There are a number of very important concepts contained in this example and they include:–

- the use of a module to define a type: for user defined data types we must create a module to define the data type if we want it to be available in more than one program unit. We will look more formally at modules in later chapters.

- the USE statement to make this data type available in each program unit that requires it: in the above case in the main program, the function and the subroutine;

- a subroutine is an extension of the facilities provided by function. It does not return a result, but carries out a more complex action. We will look at the concepts involved here in more depth in the later chapters of subroutines and modules.

- the use of interface blocks to define the interfaces to the function and subroutine. We will look at the concepts of interface blocks more fully in later chapters.

- the use of a function that returns a pointer as a result;

Dynamic Data Structures

- as the function returns a pointer we must determine the allocation status before the function terminates. This means that in the above case that we use the NULLIFY(Result) statement. The other option is to TARGET the pointer.
- the use of ASSOCIATED to determine if the node of the tree is terminated or points to another node;

Type the program in and compile, link and run it. Notice that the tree only has the minimal depth necessary to store all of the items. Experiment with the number of items and watch the tree change its depth to match the number of items. This will create a structure that is shown graphically below.

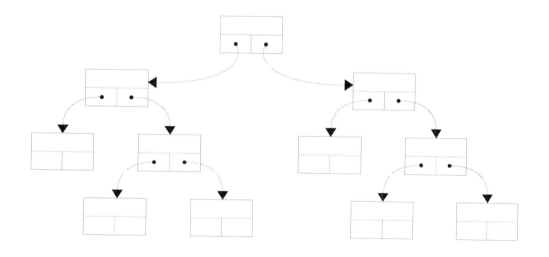

Balanced Tree

20.5 Using Linked Lists for Sparse Matrix Problems

A matrix is said to be sparse if many of its elements are zero. Mathematical models in areas such as management science, power systems analysis, circuit theory and structural analysis consist of very large sparse systems of linear equations. These systems would not be possible to solve if classical methods were applied because the sparsity would be lost and the eventual system too large to solve. Many of these systems consist of tens of thousands, hundreds of thousands and millions of equations. As the computer systems become ever more powerful with massive amounts of memory the solution of even larger problems becomes feasible.

Direct Methods for Sparse Matrices, Duff I.S. Erismon A.M., Reid J.K., looks at direct methods for solving sparse systems of linear equations.

Sparse matrix techniques lend themselves to the uses of dynamic data structures in Fortran 90. Only the non-zero elements of a sparse matrix need be stored, together with their position in the matrix. Other information also needs to be stored so that row or column manipulation can be performed without repeated scanning of a potentially very large data structure. Sparse methods may involve introducing some new non-zero elements, and a way is needed of inserting them into the data structure. This is where the Fortran 90 pointer construct can be used. The sparse matrix can be implemented using a linked list to which entries cab be easily added and deleted.

As a simple introduction consider the storage of sparse vectors. What we learn here can easily be applied to sparse matrices, which can be thought of as sets of sparse vectors.

20.5.1 Inner Product of two Sparse Vectors

Assume that we have two sparse vectors x and y and their non-zero elements are stored in sequential order in two files, together with their indices, e.g.

File 1		File 2	
i	x_i	l	y_l
j	x_j	m	y_m
k	x_k	n	y_n
.	.	.	.
.	.	.	.
.	.	.	.

and we wish to calculate the inner product $x^T y \equiv \sum_{i=1}^{n} x_i \, y_i$. There are a number of approaches to doing this and the one we use in the program below stores them as two linked lists.

```
PROGRAM C20_04
!
! This program reads the non-zero elements of two sparse
! vectors together with their indices, and stores them in
! two linked lists.
! Using these linked lists it then calculates and prints
! out the inner product. It also prints the values.
!
    IMPLICIT NONE
    CHARACTER (LEN=30):: Filename
    TYPE sparse_vector
    INTEGER :: index
    REAL:: value
    TYPE (sparse_vector), POINTER ::next
```

```
      END TYPE sparse_vector
      TYPE(sparse_vector), POINTER :: Root_x,Current_x, &
                                      Root_y,Current_y
      REAL :: Inner_prod=0.0
      INTEGER::IO_status
!
! Read first vector into a linked list
!
   PRINT *,'input file name for first vector'
   READ '(A)',Filename
   OPEN(UNIT=1,FILE=Filename,STATUS='OLD',IOSTAT=IO_status)
   IF(IO_status /= 0)THEN
      PRINT*,'Error opening file ',Filename
      STOP
   ENDIF
   ALLOCATE(Root_x)
   READ (UNIT=1,FMT=*,IOSTAT=IO_status)Root_x%value,Root_x%index
   IF(IO_status /= 0) THEN
      PRINT*,' Error when reading from file ',Filename, &
            ' or file empty'
      STOP
   ENDIF
   Current_x => Root_x
   ALLOCATE(Current_x%next)
   DO WHILE(ASSOCIATED(Current_x%next))
      Current_x => Current_x%next
      READ (UNIT=1,FMT=*,IOSTAT=IO_status)   Current_x%value, &
                                             Current_x%index
      IF(IO_status == 0)THEN
         ALLOCATE(Current_x%next)
         CYCLE
      ELSEIF(IO_status > 0 )THEN
!
! Error on reading
!
         PRINT*,'Error occurred when reading from file ',Filename
         STOP
      ELSE
!
! End of file
!
         NULLIFY (Current_x%next)
```

```fortran
      END IF
   END DO

   CLOSE(UNIT=1)
!
! Read second vector into a linked list
!
   PRINT *,'input file name for second vector'
   READ '(A)',Filename
   OPEN(UNIT=1,FILE=Filename,STATUS='OLD',IOSTAT=IO_status)
   IF(IO_status /= 0)THEN
      PRINT*,'Error opening file ',Filename
      STOP
   ENDIF
   ALLOCATE(Root_y)
   READ  (UNIT=1,FMT=*,IOSTAT=IO_status)Root_y%value,Root_y%index
   IF(IO_status /= 0) THEN
      PRINT*,' Error when reading from file ',Filename, &
             ' or file empty'
      STOP
   ENDIF
   Current_y => Root_y
   ALLOCATE(Current_y%next)
   DO WHILE(ASSOCIATED(Current_y%next))
      Current_y => Current_y%next
      READ  (UNIT=1,FMT=*,IOSTAT=IO_status)   Current_y%value, &
                                              Current_y%index

      IF(IO_status == 0)THEN
         ALLOCATE(Current_y%next)
         CYCLE
      ELSEIF(IO_status > 0 )THEN
!
! Error on reading
!
         PRINT*,'Error occurred when reading from file ',Filename
         STOP
      ELSE
!
! End of file
!
         NULLIFY (Current_y%next)
      END IF
```

```
      END DO
!
! Perform inner product
!
   Current_x => Root_x
   DO WHILE (ASSOCIATED(Current_x%next))
      Current_y => Root_y
      DO WHILE (Associated(Current_y%next) &
      .AND. Current_y%index < Current_x%index)
!
! move through 2nd list
!
         Current_y => Current_y%next
      END DO
!
! At this point Current_y%index >= Current_x%index
! or 2nd list is exhausted
!
      IF (Current_y%index == Current_x%index) THEN
         Inner_prod = Inner_prod+ Current_x%value * Current_y%value
         Current_y => Current_y%next
      END IF
      Current_x => Current_x%next
   END DO
!
! Print out inner product
!
   PRINT *,'Inner product of two sparse vectors is :', Inner_prod
!
! Print first list out
!
   PRINT*,'first list is:'
   Current_x => Root_x
   DO WHILE (ASSOCIATED(Current_x%next))
      PRINT*,Current_x%value,Current_x%index
      Current_x => Current_x%next
   END DO
!
! Print second list out
!
   PRINT*,'second list is:'
```

```
      Current_y => Root_y
      DO WHILE (ASSOCIATED(Current_y%next))
         PRINT*,Current_y%value,Current_y%index
         Current_y => Current_y%next
      END DO
!
END PROGRAM C20_04
```

20.6 Data Structures Summary

We have not been able to give examples of the wide range of data structuring techniques that exist. This is impossible in a book where the primary aim is to teach a working sub-set of a programming language, in this case Fortran 90. There are a whole range of data structures and algorithms we have not been able even to mention, e.g. finite state machines, string processing algorithms, geometric algorithms, graph algorithms, quad trees and oct trees for the manipulation of spatial information etc. Use must be made of the material in the bibliographies at the end of each chapter. The choice of the correct algorithm and data structure can make a seemingly impossible task trivial. (Well not quite trivial!).

The more you learn in this area the easier things become. It is surprising how ingenious man can be, and some of the solutions presented in the references are fascinating.

20.7 Problems

1. Type in the first example. Verify that the output is as stated, i.e.

2 2 2 4

2. There are a number of ways of handling exceptions with the READ statement, and we have used the IOSTAT option in this chapter. Consider the following program:–

```
PROGRAM C20_04
INTEGER :: IO_Stat_Number=0
INTEGER :: I
   DO
      READ (UNIT=*,FMT=10,ADVANCE='NO',IOSTAT=IO_Stat_Number) I
   10 FORMAT(I3)
! 0 = no error, no end of file (eof), no end of record (eor)
! - = eor or eof
! + = an error occurred
      PRINT *,' iostat=',IO_Stat_Number
      PRINT *,I
   END DO
END PROGRAM C20_04
```

Dynamic Data Structures

This program is a simple test of the iostat values of whatever system you work on. Try typing in a variety of values including minimally

- a valid 3 digit number + [RETURN] key
- a three digit number with an embedded blank, e.g. 1 2 + [RETURN] key
- [RETURN] key only
- [CTRL] + Z
- any other non-numeric character on the keyboard
- 100200300 + [RETURN] key
- [CTRL] + C

This will enable us to program exactly the kind of behaviour we want from i/o and can be used as a code segment for other programs.

3. Using the balanced tree example as a basis, modify it to work with a character array rather than an integer. The routine that prints the tree will also have to be modified to reflect this.

20.8 Bibliography

There are a number of books that are of use here. Most sources are Pascal and Modula 2, but there should be little difficulty converting from one of these languages to Fortran 90.

Duff I.S. Erismon A.M., Reid J.K., *Direct Methods for Sparse Matrices*, Oxford Science Publications.

- Authoritative coverage of this area. Relatively old, but well regarded. Code segments and examples are a mixture of Fortran 77 and Algol 60 (which of course do not support pointers) and therefore the implementation of linked lists is done using the existing features of these languages. The onus is on the programmer to correctly implement linked lists using fixed size arrays rather than using the features provided by pointers in a language. It is remarkable how elegant these solutions are given the lack of dynamic data structures in these two languages.

Schneider G.M., Bruell S.C., *Advanced Programming and Problem Solving with Pascal*, Wiley.

- The book is aimed at computer science students and follows the curriculum guidelines laid down in *Communications of the ACM*, August 1985, Course CS2. The book is very good for the complete beginner as the examples are very clearly laid out and well explained. There is a coverage of data structures, abstract data types and their implementation, algorithms for sorting and searching, the principles of software development as they relate to the specification, design, implementation and verification of programs in an orderly and disciplined fashion – their words.

Sedgewick, *Algorithms*, Addison Wesley.
- This book is available in a number of language flavours. The Pascal and Modula 3 books are recommended.

Wirth N., *Algorithms + Data Structures = Programs*, Prentice Hall
- An early but illuminating book on the subject. Well worth a read. Pascal is used.

21
Files

It is a capital mistake to theorise before one has data.
Sir Arthur Conan Doyle.

Aims

The aims of this chapter are:–

- to review the process of file creation at a terminal
- to introduce more formally the idea of the file as a fundamental entity
- to show how files can be declared explicitly by the OPEN and CLOSE statements
- to introduce the arguments for the OPEN and CLOSE statements
- to demonstrate the interaction between the READ/WRITE statements and the OPEN/CLOSE statements;

21 Files

While you are working interactively, on a terminal, you will be working with files; files that contain programs, files that contain data, and perhaps files that are libraries. The file is fundamental to most modern operating systems, and almost all operations are carried out on files.

In this chapter we are going to extend some of your ideas about files. Let us consider what kinds of files you have met so far:–

1) Text files. These are the source of your programs, compilation listings etc. They can be examined by printing them. They can also be transmitted round a computer system fairly easily. A file sent to a printer is a text file. Mail messages are generally plain text files. Note that when mail messages arrive in your mail box they will then typically contain additional non-printable information.

2) Data files. These exist in two main forms, firstly those prepared by using an editor, and hence a text file, and those prepared using a package or program, and in a computer readable form, but not directly readable by a human.

3) Binary, object or relocatable files, e.g. output from the compiler, satellite data. They cannot be printed. To examine files like these you need to use special utilities, provided by most operating systems.

The above categories account for the majority of files that you have met so far.

If you use a word processor then you will also have met files that are textual with additional non-printable information.

Let us now consider how we can manipulate files using Fortran. They will generally be data files, and will thus be text files. They can therefore be listed etc., using standard operating system commands.

21.1 Files in Fortran

These allow us to associate a logical unit number with any arbitrary file name during the running of the program, e.g.

```
OPEN(UNIT=1,FILE='DATA')
```

would associate the name DATA and the logical unit 1, so that

```
READ(UNIT=1,FMT=100) X
```

would read from DATA. Note that for this to work on some operating systems the file DATA must have been 'local' to the session; we specify the name as a character variable. If we then wanted to use a subsequent data file, we could have another OPEN statement, but if we want to use the same logical unit number, we must first CLOSE the file.

Files

```
CLOSE(UNIT=1,FILE='DATA')
```

before we

```
OPEN(UNIT=1,FILE='DATA2')
```

In this way we can keep referring to logical unit 1, but change the file associated with it. This can be useful in interactive programs where we wish to analyse different sets of data, e.g.

```
PROGRAM Flex
IMPLICIT NONE
REAL :: X
CHARACTER (7) :: WHICH
   OPEN(UNIT=5,FILE='INPUT')
   DO
      WRITE(UNIT=6,FMT='('' DATA SET NAME, OR END'')')
      READ(UNIT=5,FMT='(A)') WHICH
      IF(WHICH == 'END') EXIT
      OPEN(UNIT=1,FILE=WHICH)
      READ(UNIT=1,FMT=100) X
      ...
      CLOSE(UNIT=1,FILE=WHICH)
   END DO
END PROGRAM Flex
```

One useful feature of the OPEN statement is that there are other parameters. What would happen, for example, if the file was not there? To take care of this you can use the IOSTAT and STATUS keywords, e.g.

```
OPEN(UNIT=1,FILE='DATA',IOSTAT=Open_File_Status,STATUS='OLD')
```

STATUS can be equated to one of four values:–

```
STATUS='OLD'
STATUS='NEW'
STATUS='SCRATCH'
STATUS='UNKNOWN'
```

If we say STATUS='NEW', we are creating a new file and it should not matter whether a file of the same name is present; 'SCRATCH' does not concern us, while 'UNKNOWN' implies that, if a file of the correct name is present use it, if not create a 'NEW' one. If you omit the STATUS= keyword altogether, the value 'UNKNOWN' will be assumed. If we use STATUS='OLD' and the file is not present, this will cause an error and this will reflected in the value associated with the variable Open_File_Status, consider the following example:–

```
...
OPEN(UNIT=1,FILE='DATA',IOSTAT=Open_File_Status,STATUS='OLD')
IF (Open_File_Status > 0) THEN
   PRINT *,' Error opening file, please check'
   STOP
END IF
READ(UNIT=1,FMT=100) X
...
```

The program will terminate after printing an appropriate error message. The standard defines that if an error occurs then IOSTAT will return a positive integer value. A value of zero is returned if no error occurs.

21.2 Summary of options on OPEN

UNIT The unit number of the file to be opened.

IOSTAT Integer variable given the value zero if there are no errors.

FILE Character expression specifying the file name.

STATUS Character expression specifying the file status. It can be one of 'OLD', 'NEW', 'SCRATCH' or 'UNKNOWN'.

ACCESS Character expression specifying whether the file is to be used in a sequential or random fashion. Valid values are *SEQUENTIAL* (the default), or *DIRECT*.

The two most common access mechanisms for files are sequential and direct. Consider a file with 1000 records. To get at record 789 in a sequential file means reading or processing the first 788 records. To get at record 789 in a direct access file means using a record number to immediately locate record 789.

FORM Character expression specifying one of:–

FORMATTED if the file is opened for formatted i/o.

UNFORMATTED if the file is opened for unformatted i/o.

The default is formatted for sequential access files and unformatted for direct access files. If the file exists, FORM must be consistent with its present characteristics.

As was stated in earlier chapters data is maintained internally in a binary format, not immediately comprehensible by humans. When we wish to look at the data we must write it out in a formatted fashion, i.e. as a sequence of printable ASCII characters – text, or the written word. This formatting will carry with it an overhead in terms of the time required to do the formatting. It will also carry with it the penalty of conversion from one number base (internally binary) to another and also loss of significance due to rounding with whatever edit descriptors are used, e.g. writing out as F7.4.

Files

If we are interested in reusing data on the same system and compiler then we can use the unformatted option and avoid both the time overhead (as there is no conversion between the internal and external formats) and the loss of significance associated with formatted data.

Please note that unformatted files are rarely portable between different computer systems, and sometimes even between different compilers on the same system.

We will look again at the use of unformatted files in chapter 28 when we look at efficiency and the space time trade off.

RECL Integer variable or constant specifying the record length for a direct access file. It is specified in characters for a formatted file, and words for an unformatted file.

BLANK Character expression having one of the following values:–

'NULL' if blanks are to be ignored on reading. Note that a field of all blanks is treated as 0!

'ZERO' if blanks are to be treated as zeros.

21.3 More fool proof i/o

Fortran provides a way of writing more fool proof programs involving i/o. This is done via the IOSTAT keyword on the READ statement. Consider the program given as a problem at the end of chapter 20:–

```
PROGRAM C21_01
IMPLICIT NONE
INTEGER :: IO_Stat_Number=0
INTEGER :: I
   DO
      READ (UNIT=*,FMT=10,ADVANCE='NO',IOSTAT=IO_Stat_Number) I
      10 FORMAT(I3)
      PRINT *,' iostat=',IO_Stat_Number
      PRINT *,I
   END DO
END PROGRAM C21_01
```

If you didn't attempt this problem then the above program should be typed in and run on whatever system you use. There are aspects of it that are implementation specific. The following data input should be tried and the values of IO_Stat_Number should be examined:–

1. a valid 3 digit number + [RETURN] key
2. a three digit number with an embedded blank, e.g. 1 2 + [RETURN] key
3. [RETURN] key only
4. [CTRL] + Z
5. any other non-numeric character on the keyboard

6. 100200300 + [RETURN] key=
7. [CTRL] + C

This will then enable us to write programs that handle common i/o errors. Consider the following:–

```
PROGRAM C21_02
INTEGER , DIMENSION(10) :: A
INTEGER :: IO_Stat_Number=0
INTEGER :: I
   OPEN(UNIT=1,FILE='DATA.DAT',STATUS='OLD')
   DO I=1,10
      READ (UNIT=1,FMT=10,IOSTAT=IO_Stat_Number) A(I)
      10 FORMAT(I3)
      IF (IO_Stat_Number == 0) THEN
         CYCLE
      ELSEIF (IO_Stat_Number == -2) THEN
         PRINT *,' End of record detected at line ',I
         PRINT *,' Please check data file'
         EXIT
      ELSEIF (IO_Stat_Number == -1) THEN
         PRINT *,' End of file detected at line ',I
         PRINT *,' Please check data file'
         EXIT
      ELSEIF (IO_Stat_Number > 0 ) THEN
         PRINT *,' Non numeric data in file at line ',I
         PRINT *,' Please correct data file'
         EXIT
      ENDIF
   END DO
   .
   .
   .
END PROGRAM C21_02
```

The above program is system specific but interestingly both the NAG/Salford compiler and the DEC Alpha OPENVMS compiler return the same values for end of record and end of file.

Note that in the above example the testing for the various conditions only exits the DO loop for reading data from the file. This means that execution would continue with the statement immediately after the END DO statement. This may not be what we want in all cases, and the EXIT may be replaced with a STOP statement to terminate execution immediately.

Files

21.4 Summary

• The file is a fundamental entity within the operating system.

• Files may be manipulated in Fortran by associating their name with a unit number. All subsequent communication within the program is through the unit number.

• When a file is opened there are a large number of equatable keywords which may be employed to establish its characteristics.

• The default file type used in Fortran is *sequential formatted,* but several other esoteric types may be used.

21.5 Problems

1. Write a program to write the first 500 integers to a file using formatted i/o. Put 10 values on a line, with a blank as the first character of the line, and 8 columns allowed for each integer, with two spaces between each integer field.

Now write a program to read this file into an array, and write the numbers in reverse order over the original data. i.e. the data file now contains the first 500 numbers in descending order.

Now modify the first program to add the next 500 integers to the same file, so that the file now comprises the first 500 numbers in descending order, and the next 500 numbers in ascending order.

2. To write and maintain a crude data base of student details, we might do the following; create separate files for each year — CLAS1, CLAS2, CLAS3, or COF84, COF85, COF86, and so on. In either case there is an unchanging prefix, CLAS or COF, and a variable suffix, which identifies membership within the overall group. On each of the files we may wish to record details like; name, date of birth, address, courses taken, etc. Such files will require updating as details change, or as errors are noted. Write (or sketch out) a program which would select and maintain such records, and which would allow corrected files to be printed out. While you might feel that the most appropriate tool for this job is an editor, you might find this too powerful a tool. An editor can leave files in a sorry state. Naturally, any program like this should be helpful (so called 'user friendly'). Is this sort of information sensitive enough to require security checks and passwords?

22
Introduction to Subroutines

A man should keep his brain attic stacked with all the furniture he is likely to use, and the rest he can put away in the lumber room of his library, where he can get at it if he wants.

Sir Arthur Conan-Doyle, *Five Orange Pips*.

Aims

The aims of this chapter are:–

- to consider some of the reasons for the inclusion of subroutines in a programming language;
- to introduce with a concrete example some of the concepts and ideas involved with the definition and use of subroutines:–
  - the INTERFACE statement and interface blocks;
  - parameters;
  - the INTENT attribute for parameters;
  - the CALL statement;
  - scope of variables;
  - local variables and local variables and the SAVE attribute;
  - the use of parameters to report on the status of the action carried out in the subroutine;

22 An Introduction to Subroutines

In the earlier chapter on functions we introduced two types of function:–

- intrinsic functions – which are part of the language
- user defined functions – by which we extend the language

We now introduce subroutines which collectively with functions are given the name procedures. Procedures provide a very powerful extension to the language by:

- providing us with the ability to break problems down into simpler more easily solvable sub-problems.
- allowing us to concentrate on one aspect of a problem at a time.
- avoiding duplication of code.
- hiding away messy code so that a main program is a sequence of calls to procedures.
- providing us with the ability to put together collections of procedures that solve commonly occurring sub-problems, often given the name libraries, and generally compiled.
- allowing us to call procedures from libraries written, tested and documented by experts in a particular field. There is no point in re-inventing the wheel!

There are a number of concepts required for the successful use of subroutines and we met some of them in the chapter on functions when we looked at user defined functions. We will extend the ideas introduced there of parameters and introduce the additional concept of an interface block. The ideas are best explained with a concrete example.

Note that we use the terms parameters and arguments interchangeably.

22.1 Simple Subroutine Example

This example is one we met earlier that solves a quadratic equation, i.e. solves $ax^2 + bx + c = 0$

The program to do this originally was just one program. In the example below we break that problem down into smaller parts, and make each part a subroutine. The components are

- main program or driving routine;
- interaction with user to get the coefficients of the equation;
- solution of the quadratic;

Let us look now at how we do this with the use of subroutines.

```
PROGRAM C22_01
IMPLICIT NONE
! Simple example of the use of a main program and two
! subroutines. One interacts with the user and the second
! solves a quadratic equation, based on the use input.
INTERFACE

   SUBROUTINE Interact(A,B,C,OK)
      IMPLICIT NONE
      REAL , INTENT(OUT) :: A
      REAL , INTENT(OUT) :: B
      REAL , INTENT(OUT) :: C
      LOGICAL , INTENT(OUT) :: OK
   END SUBROUTINE Interact

   SUBROUTINE Solve(X,Y,Z,Root1,Root2,IFail)
      IMPLICIT NONE
      REAL , INTENT(IN) :: X
      REAL , INTENT(IN) :: Y
      REAL , INTENT(IN) :: Z
      REAL , INTENT(OUT) :: Root1
      REAL , INTENT(OUT) :: Root2
      INTEGER , INTENT(INOUT) :: IFail
   END SUBROUTINE Solve

END INTERFACE

REAL :: P, Q, R, Root1, Root2
INTEGER :: IFail=0
LOGICAL :: OK=.TRUE.
   CALL Interact(P,Q,R,OK)
   IF (OK) THEN
      CALL Solve(P,Q,R,Root1,Root2,IFail)
      IF (IFail == 1) THEN
         PRINT *,' Complex roots, calculation abandoned'
      ELSE
         PRINT *,' Roots are  ',Root1,'   ',Root2
      ENDIF
   ELSE
      PRINT*,' Error in data input program ends'
   ENDIF
END PROGRAM C22_01
```

An Introduction to Subroutines

```fortran
SUBROUTINE Interact(A,B,C,OK)
   IMPLICIT NONE
   REAL , INTENT(OUT) :: A
   REAL , INTENT(OUT) :: B
   REAL , INTENT(OUT) :: C
   LOGICAL , INTENT(OUT) :: OK
   INTEGER :: IO_Status=0
   PRINT*,' Type in the coefficients A, B AND C'
   READ(UNIT=*,FMT=*,IOSTAT=IO_Status)A,B,C
   IF (IO_Status == 0) THEN
      OK=.TRUE.
   ELSE
      OK=.FALSE.
   ENDIF
END SUBROUTINE Interact

SUBROUTINE Solve(X,Y,Z,Root1,Root2,IFail)
   IMPLICIT NONE
   REAL , INTENT(IN) :: X
   REAL , INTENT(IN) :: Y
   REAL , INTENT(IN) :: Z
   REAL , INTENT(OUT) :: Root1
   REAL , INTENT(OUT) :: Root2
   INTEGER , INTENT(INOUT) :: IFail
! Local variables
   REAL :: Term
   REAL :: A2
   Term = Y*Y - 4.*X*Z
   A2 = X*2.0
! if term < 0, roots are complex
   IF(Term < 0.0)THEN
      IFail=1
   ELSE
      Term = SQRT(Term)
      Root1 = (-Y+Term)/A2
      Root2 = (-Y-Term)/A2
   ENDIF
END SUBROUTINE Solve
```

22.2 Defining a subroutine

A subroutine is defined as:–

>   SUBROUTINE subroutine_name(optional list of dummy arguments)
>
>   >   dummy argument type definitions with INTENT
>
>   >   ...
>
>   END SUBROUTINE subroutine_name

and from the earlier example we have the subroutine:–

```
SUBROUTINE Interact(A,B,C,OK)
   REAL,    INTENT(OUT)::A,B,C
   LOGICAL, INTENT(OUT)::OK


END SUBROUTINE Interact
```

22.3 Referencing a subroutine

To reference a subroutine you use the CALL statement:–

CALL subroutine_name(optional list of actual arguments)

and from the earlier example the call to subroutine Interact was of the form:–

```
CALL Interact(P,Q,R,OK)
```

When a subroutine returns to the calling program unit control is passed to the statement following the CALL statement.

22.4 Dummy Arguments or Parameters, and Actual Arguments

Procedures and their calling program units communicate through their arguments. We often use the terms parameter and arguments interchangeably throughout this text. The SUBROUTINE statement normally contains a list of dummy arguments, separated by commas and enclosed in brackets. The dummy arguments have a type associated with them, for example in subroutine Solve X is of type REAL, but no space is put aside for this in memory. When the subroutine is referenced e.g. CALL Solve(P,Q,R,Root1,Root2,Ifail) then the dummy argument *points* to the actual argument P which is a variable in the calling program unit. The dummy argument and the actual argument must be of the same type, in this case REAL.

22.5 Interface

We strongly recommend the use of interface blocks in the calling routine to provide us with information about the called routine. These provide us with the ability to do type checking between the

An Introduction to Subroutines

calling routine and the called routine. One of the most common errors in programming is getting the sequence and type of the parameters wrong between sub-programs.

There are times when the use of interface blocks is mandatory in Fortran and we will cover this as and when required. However it is a good working practice to always provide interface blocks.

```
INTERFACE

    SUBROUTINE Interact(A,B,C,OK)
      IMPLICIT NONE
      REAL  , INTENT(OUT) :: A
      REAL  , INTENT(OUT) :: B
      REAL  , INTENT(OUT) :: C
      LOGICAL , INTENT(OUT) :: OK
    END SUBROUTINE Interact

    SUBROUTINE Solve(X,Y,Z,Root1,Root2,IFail)
      IMPLICIT NONE
      REAL  , INTENT(IN) :: X
      REAL  , INTENT(IN) :: Y
      REAL  , INTENT(IN) :: Z
      REAL  , INTENT(OUT) :: Root1
      REAL  , INTENT(OUT) :: Root2
      INTEGER , INTENT(INOUT) :: IFail
    END SUBROUTINE Solve

END INTERFACE
```

The subroutine Interact takes four arguments, A, B, C and OK, and the first three are of REAL type and the fourth is of type LOGICAL.

The subroutine Solve takes six arguments, X,Y,Y, Root1, Root2 and IFail and the first five are of REAL type and the sixth is of INTEGER type.

As Fortran 90 libraries become more widely available interface blocks for library routines will be provided by the supplier on line, and this minimises much of the effort in using them. NAG for example already has interface blocks available for its library for many platforms.

22.6 Intent

It is recommended that dummy arguments have an INTENT attribute. In the earlier example subroutine Solve has a dummy argument X with INTENT(IN), which means that when the subroutine is referenced or called it is expecting X to have a value , but its value cannot be changed inside the subroutine. This acts as an extra security measure besides making the program easier to understand. For each parameter it may have one of three attributes:–

- INTENT(IN), where the parameter already has a value and cannot be altered in the called routine;
- INTENT(OUT), where the parameter does not have a value, and is given one in the called routine;
- INTENT(INOUT), where the parameter already has a value and this is changed in the called routine;

22.7 Local Variables

We saw with functions that variables could be essentially local to the function and unavailable elsewhere. The concept of local variables also applies to subroutines. In the example above Term and A2 are both local variables to the subroutine Solve.

22.7.1 Local Variables and the SAVE attribute

Local variables are usually created when a procedure is called and their value lost when execution returns to the calling program unit. To make sure that a local variable retains its values between calls to a sub-program the SAVE attribute can be used on a type statement, e.g.

```
INTEGER , SAVE :: I
```

means that when this statement appears in a sub-program the value of the local variable I is saved between calls.

22.8 Scope of Variables

In most cases variables are only available within the program unit that defines them. The introduction of argument lists to functions and subroutines immediately opens up the possibility of data within one program unit becoming available in one or more program units.

In the main program we declare the variables P, Q, R, Root1, Root2, IFail and OK.

Subroutine Interact has no variables locally declared. It works on the arguments A, B, C and OK which map onto P, Q, R and OK from the main program, i.e. it works with those variables.

Subroutine Solve has two locally defined variables Term and A2. It works with the variables X,Y,Z,Root1,Root2 and IFail, which map onto P,Q,R,Root1,Root2 and IFail from the main program.

22.9 Status of the Action Carried out in the Subroutine

It is also useful to use parameters that carry information regarding the status of the action carried out by the subroutine. With the subroutine Interact we use a logical variable OK to report on the status of the interaction with the user. In the subroutine Solve we use the status of the integer variable Ifail to report on the status of the solution of the equation.

An Introduction to Subroutines

22.10 Why Bother?

Given the increase in the complexity of the overall program to solve a relatively straightforward problem one must ask why bother. The answer lies in our ability to manage the solution of larger and larger problems. We need all the help we can get if we are to succeed in our task of developing large scale reliable programs.

We need to be able to break our problems down into manageable sub-components and solve each in turn. We are now in a very good position to be able to do this. Given a problem that requires a main program, one or more functions and one or more subroutines we can work on each sub-component in relative isolation, and know that by using things like interface blocks that we will be able to glue each of the components together into a stable structure at the end. We can independently compile the main program and functions and subroutines and use the linker to generate the overall executable and then test that. Providing we keep our interfaces the same we can alter the actual implementations of the functions and subroutines and just recompile the changed procedures.

22.11 Summary

We now have the following concepts for the use of subroutines:–

- INTERFACE blocks;
- INTENT attribute for parameters;
- Dummy parameters;
- the use of the CALL statement to invoke a subroutine;
- the concepts of variables that are local to the called routines and are unavailable elsewhere in the overall program;
- communication between program units via the argument list;
- the concept of parameters on the call that enable us to report back on the status of the called routine;

22.12 Problems

1. Type in the program example in this chapter as three files. Compile each individually. When you have successfully compiled each routine (there will be the inevitable typing mistakes) look at the file sizes of the object file. Now use the linker to produce one executable. Look at the file size of the executable. What do you notice?

The development of large programs is eased considerably by the ability to compile small program units and eradicate the compilation errors from one unit at a time.

The linker obviously also has an important role to play in the development process.

23
Subroutines: 2

It is one thing to show a man he is in error, and another to put him possession of the truth.

John Locke

Aims

The aims of this chapter are to extend the ideas in the earlier chapter on subroutines by using concrete examples that:–

- look at passing rank one arrays as parameters, rather than simple scalars;
    - to introduce explicit shape dummy arrays;
- look at passing character variables as parameters;
    - to introduce assumed length dummy arguments;
- uses the Quick Sort algorithm which in turn;
    - introduces recursive subroutines;
    - introduces the use of an internal subroutine;
    - looks at the way in which we can maintain a consistent interface between a program and a subroutine and actually re-implement the algorithm that solves the particular sub-problem;
    - looks at the use of the timing routines available in Fortran to determine where the program actually spends its time;
- look at passing rank two arrays as parameters;
    - to introduce assumed shape dummy arrays;

23 Subroutines: 2

23.1 Example 1: Introduction to Arrays as Parameters

23.1.1 Explicit Shape Dummy Arrays

So far we have seen scalar parameters. We will now look at array parameters, initially rank one arrays.

A fundamental rule in Fortran 90 is that the shape of an actual argument and its associated dummy arguments are the same, i.e., they must both have the same rank and the same extents in each dimension.

Consider a simple program which calls a subroutine to add up the square and the cube of each element in a real array, to produce:–

$$\sum_{i=1}^{n} x_i^2 \quad \text{and} \quad \sum_{i=1}^{n} x_i^3$$

```
PROGRAM Ch23_1
IMPLICIT NONE
REAL, DIMENSION (1:50)::A
REAL:: Sum_A_2,Sum_A_3
INTEGER::Number
INTERFACE
   SUBROUTINE Stats(X,Sum_x_2,Sum_x_3,N)
      IMPLICIT NONE
      REAL,INTENT(IN),DIMENSION(1:50)::X
      REAL,INTENT(OUT)::Sum_x_2,Sum_x_3
      INTEGER,INTENT(IN)::N
   END SUBROUTINE Stats
END INTERFACE
   ...
   CALL Stats(A,Sum_A_2,Sum_A_3,Number)
   ...
END PROGRAM Ch23_1

SUBROUTINE Stats(X,Sum_x_2,Sum_x_3,N)
IMPLICIT NONE
REAL,INTENT(IN),DIMENSION(1:50)::X
REAL,INTENT(OUT)::Sum_x_2,Sum_x_3
INTEGER,INTENT(IN)::N
INTEGER::I
REAL:: Temp
```

```
!
   Sum_x_2=0.0
   Sum_x_3=0.0
   DO I=1,N
      Temp=X(I)*X(I)
      Sum_x_2=Sum_x_2+Temp
      Sum_x_3=Sum_x_3+X(I)*Temp
   END DO
END SUBROUTINE Stats
```

In both the subroutine and the main program the dummy argument X and the actual argument A are the same shape. In subroutine Stats the dummy argument X is declared with explicit bounds, and is known as an explicit shape dummy array. This program and subroutine will work with arrays of up to and including 50 elements. To allow for larger problems the upper bounds of the actual and dummy array arguments must be increased. This doesn't make the program and subroutine very general. One way round this is to give the actual argument a deferred shape by making it an allocatable array in the main program and to make the dummy argument N the upper bound for the dummy array X in the subroutine, e.g. in the main program use:–

```
REAL, DIMENSION (:), ALLOCATABLE:: A
   READ *, Number
   ALLOCATE(A(1:Number))
   CALL Stats(A,Sum_A_2,Sum_A_3,Number)
```

and in the subroutine use:-

```
REAL, INTENT(IN), DIMENSION(1:N) ::X
```

The dummy argument X still has explicit shape, but now the bounds depend on the value of the dummy argument N and its size is determined when the subroutine is called. Even though it is not mandatory it is recommended that interface blocks are used so that the shape and size of actual and dummy arguments can be checked explicitly.

23.2 Example 2: Characters as parameters and assumed length dummy arguments

The types of parameters considered so far have been REAL, INTEGER and LOGICAL. CHARACTER variables are slightly different because they have a length associated with them. In the simplest form consider a subroutine which given the name of a file, opens it and reads values into two REAL arrays X and Y.

```
SUBROUTINE Readin(Name,X,Y,N)
IMPLICIT NONE
REAL,DIMENSION(1:N),INTENT(OUT)::X,Y
CHARACTER (LEN=20),INTENT(IN)::Name
```

```
INTEGER::I
   OPEN(UNIT=10,STATUS='OLD',FILE=Name)
   DO I=1,N
      READ(10,*)X(I),Y(I)
   END DO
   CLOSE(UNIT=10)
END SUBROUTINE Readin
```

A main program to call this subroutine could look like the following:–

```
PROGRAM Ch23_2
IMPLICIT NONE
REAL,DIMENSION(1:100)::A,B
INTEGER::Nos
CHARACTER(LEN=20)::Filename
INTERFACE
   SUBROUTINE Readin(Name,X,Y,N)
      IMPLICIT NONE
      REAL,DIMENSION(1:N),INTENT(OUT)::X,Y
      CHARACTER (LEN=20),INTENT(IN)::Name
   END SUBROUTINE Readin
END INTERFACE
   ...
   CALL Readin(Filename,A,B,Nos)
   ...
END PROGRAM Ch23_2
```

The main program reads the file name from the user and passes it to the subroutine that reads in the data. If we wanted to allow longer filenames then we would have to change both the main program and the subroutine. To minimise the number of places where we have to make changes we can use assumed length dummy arguments and their use is strongly recommended for character arguments. The type statement in the subroutine should be therefore changed to

```
CHARACTER (LEN=*) , INTENT(IN) :: Name
```

and the dummy argument Name then takes on the size of the actual argument Filename. An interface block **must** be used when we are using assumed shape dummy arguments.

23.3 Example 3: Using Hoare's Quick Sort Algorithm

This example highlights several important issues involved in the use of subroutines. The overall problem is to read in and sort a file of real numbers. This can be broken down into several steps:–

- interact with the user and find out the file name and how many items there are in the file;

- sort these items
- write out the sorted data to a second file.

We will add an additional step of producing timing information regarding each subroutine in the overall problem solution.

We will solve the problem by writing the following routines:–

- a driving routine or main program with calls to the timing routines and a print out of the times spent in each subroutine. We will use the intrinsic subroutine SYSTEM_CLOCK to produce the timing information;
- a subroutine to read the data from a file;
- a subroutine to sort the data;
- a subroutine to write the data to a second file.

In the problem at the end we will look at replacing the actual routine that does the sort with another routine. If you have the NAG library or similar available, use one of the sort routines in that library, and see what timing that routine provides.

```
PROGRAM Sort_Example01

! This program is a simple example of how to construct larger programs
! using some of the new features introduced with Fortran 90. The program
! comprises 4 routines,
! 1. a driving routine, which interacts with the user
! and requests the number of data items, and file name.
! 2. a routine to read in the data, from a file.
! 3. a routine to sort the data.
! 4. a routine to print out the sorted data to a file
!
! The program attempts to highlight the importance of the
! modular nature of the solution, and how we can simply
! replace one component, without affecting the other routines.
IMPLICIT NONE
INTEGER                                  :: How_Many
CHARACTER    (LEN=20)                    :: File_Name
REAL , ALLOCATABLE , DIMENSION(:)        :: Raw_Data
INTEGER                                  :: Count
INTEGER                                  :: Count_Rate
INTEGER                                  :: T1,T2,T3,T4,I,J
```

Subroutines: 2

```fortran
      INTERFACE
         SUBROUTINE Read_Data(File_Name,Raw_Data,How_Many)
            IMPLICIT NONE
            CHARACTER (LEN=*) , INTENT(IN) :: File_Name
            INTEGER , INTENT(IN) :: How_Many
            REAL , INTENT(OUT) , DIMENSION(How_Many) :: Raw_Data
         END SUBROUTINE Read_Data
      END INTERFACE

      INTERFACE
         SUBROUTINE Sort_Data(Raw_Data,How_Many)
            IMPLICIT NONE
            INTEGER , INTENT(IN) :: How_Many
            REAL , INTENT(INOUT) , DIMENSION(How_Many) :: Raw_Data
         END SUBROUTINE Sort_Data
      END INTERFACE

      INTERFACE
         SUBROUTINE Print_Data(Raw_Data,How_Many)
            IMPLICIT NONE
            INTEGER , INTENT(IN) :: How_Many
            REAL , INTENT(IN) , DIMENSION(How_Many) :: Raw_Data
         END SUBROUTINE Print_Data
      END INTERFACE

      PRINT * , ' How many data items are there?'
      READ  * , How_Many
      PRINT * , ' What is the file name?'
      READ '(A)',File_Name

      CALL SYSTEM_CLOCK(Count,Count_Rate,Count_Max)
      I=Count
      J=I
      PRINT *,I,'   ',Count_Rate,'   ',Count_Max

      ALLOCATE(Raw_Data(How_Many))

      CALL SYSTEM_CLOCK(Count)
      T1=Count-I
      I=T1/Count_Rate
      PRINT *,T1,I
```

```fortran
      CALL Read_Data(File_Name,Raw_Data,How_Many)

      CALL SYSTEM_CLOCK(Count)
      T2=Count-T1-J
      I=T2/Count_Rate
      PRINT *,T2,'   ',I

      CALL Sort_Data(Raw_Data,How_Many)

      CALL SYSTEM_CLOCK(Count)
      T3=Count-T2-J
      I=T3/Count_Rate

      PRINT *,T3,'   ',I

      CALL Print_Data(Raw_Data,How_Many)

      CALL SYSTEM_CLOCK(Count)
      T4=Count-T3-J
      I=T4/Count_Rate
      PRINT *,T4,'   ',I
      PRINT *,T1,'   ',T2,'   ',T3,'   ',T4
      PRINT *,' Program ends, data written to file name SORTED.DAT'

END PROGRAM Sort_Example01

SUBROUTINE Read_Data(File_Name,Raw_Data,How_Many)
IMPLICIT NONE
CHARACTER (LEN=*) , INTENT(IN) :: File_Name
INTEGER , INTENT(IN) :: How_Many
REAL , INTENT(OUT) , DIMENSION(How_Many) :: Raw_Data

INTEGER :: I

   OPEN(FILE=File_Name,UNIT=1)
   DO I=1,How_Many
      READ (UNIT=1,FMT=*) Raw_Data(I)
   ENDDO

END SUBROUTINE Read_Data
```

Subroutines: 2

```fortran
   SUBROUTINE Sort_Data(Raw_Data,How_Many)
   IMPLICIT NONE
   INTEGER , INTENT(IN) :: How_Many
   REAL , INTENT(INOUT) , DIMENSION(How_Many) :: Raw_Data
      CALL QuickSort(1,How_Many)

CONTAINS

      RECURSIVE SUBROUTINE QuickSort(L,R)
      IMPLICIT NONE
      INTEGER , INTENT(IN) :: L,R
      INTEGER :: I,J
      REAL :: V,T

         IF (R > L) THEN
            V=Raw_Data(R)
            I=L-1
            J=R
            DO
               DO
                  I=I+1
                  IF (Raw_Data(I) >= V) EXIT
               END DO
               DO
                  J=J-1
                  IF (Raw_Data(J) <= V) EXIT
               END DO
               T=Raw_Data(I)
               Raw_Data(I)=Raw_Data(J)
               Raw_Data(J)=T
               IF (J <= I) EXIT
            END DO
            Raw_Data(J)=Raw_Data(I)
            Raw_Data(I)=Raw_Data(R)
            Raw_Data(R)=T
            CALL QuickSort(L,I-1)
            CALL QuickSort(I+1,R)
         END IF
      END SUBROUTINE QuickSort

END SUBROUTINE Sort_Data
```

```
SUBROUTINE Print_Data(Raw_Data,How_Many)
IMPLICIT NONE
INTEGER , INTENT(IN) :: How_Many
REAL , INTENT(IN) , DIMENSION(How_Many) :: Raw_Data
INTEGER :: I
   OPEN(FILE='SORTED.DAT',UNIT=2)
   DO I=1,How_Many
      WRITE(UNIT=2,FMT=*) Raw_Data(I)
   ENDDO
   CLOSE(2)
END SUBROUTINE Print_Data
```

Note 1 – Interface Blocks

Each subroutine has an interface block in the main program. This enables us to compile the main program and each subroutine independently and have the benefit of cross routine type checking for each of the parameters. In the earlier chapter the parameters were scalars, in this example we have a mix of arrays and scalars.

It is good programming practice to get in the habit of using interface blocks when writing programs that contain several procedures, i.e. either functions or subroutines. The likelihood of errors decreases with the use of interface blocks.

Interface blocks are **mandatory** under the following situations:–

- the procedure has optional arguments. The intrinsic subroutine SYSTEM_CLOCK is used in this example with two arguments initially, and one subsequently, as we need only determine the Count_Rate once;
- when a function returns an array;
- when a function returns a pointer. The balanced tree example in chapter 20 uses a recursive function that returns a pointer;
- for character functions a result that is dynamic;
- when the procedure has assumed shape dummy arguments;
- when the procedure has dummy arguments with the pointer attribute;
- when the procedure has dummy arguments with the target attribute;
- when the procedure has keyword arguments and/or optional arguments;
- when the procedure is generic. You have already seen that some of the intrinsic procedures have this status, e.g. SINE will return a result when the argument is of a variety of types. This means that we can construct procedures that will accept arguments of a variety of types and all we need to do is provide a procedure that manipulates data of that type. We will look at the construction of a procedure that is generic in a later chapter;

- when the procedure provides a user defined operator;
- when the procedure provides user defined assignment.

23.3.1 Note 2 – Intent Attribute

There is the use of the INTENT attribute for each parameter. This clarifies the actions of the various routines, and decreases the likelihood of a mistake.

23.3.2 Note 3 – Explicit shape dummy array

An explicit shape dummy array is used for the data array. Its upper bound is passed as a dummy argument and is hence used to dimension the Read_Data array.

23.3.3 Note 4 – Assumed Length Dummy Argument

The subroutine Read_Data has an assumed length dummy character argument through which the file name is passed.

23.3.4 Note 5 – Recursive Subroutine

The actual sorting is done in the recursive subroutine QuickSort. This is a direct translation of Hoare's original algorithm. Recursion is said to be inefficient. In the light of the timing figures do you think that that statement is true?

Recursion provides us with a very clean and expressive way of solving many problems. There will be instances where it is worthwhile removing the overhead of recursion, but the first priority is the production of a program that is correct. It is pointless having a very efficient but incorrect problem solution.

We will look again at recursion and efficiency in a later chapter and under what criteria we can replace recursion with iteration.

23.3.5 Note 6 – Internal Subroutines and Scope

The routine that actually performs the sort is an internally declared subroutine. We first came across internal sub programs in the chapter on functions, using Stirling's approximation. Let us look at the complete program now and consider fully the whole issue of the scope of variables.

In the main program we declare the variables How_Many, File_Name, Raw_Data, Count, Count_Rate, T1, T2, T3, T4, I and J.

How_Many, File_Name and Raw_Data are arguments to Read_Data and are therefore available to that routine.

How_Many and Raw_Data are arguments to both Sort_Data and Print_Data and are therefore available to those two routines.

Count and Count_Rate are arguments to the first call to SYSTEM_CLOCK and are thus available to that routine.

Count is then an argument to subsequent calls to SYSTEM_CLOCK. Thus Count is made available to each of these calls.

T1, T2, T3, T4, I and J are purely local to the main program.

Subroutine Read_Data has one locally declared variable I, and this exists completely independently of I in the main routine.

Subroutine Print_Data has one locally declared variable I, and again this exists completely independently of I in both the main program and the Read_Data routine.

Subroutine Sort_Data and the recursive subroutine QuickSort introduce a number of important extensions to the concept of scope.

Within the Sort_Data routine the Raw_Data array is a variable that is available to the QuickSort subroutine as QuickSort is **contained** within the Sort_Data routine.

QuickSort has its own local variables I,J,V and T. Note that as QuickSort is recursive that there will be multiple versions of this routine in existence as the sorting is undertaken. Thus there will be multiple instances of the local variables I,J,V and T in existence, and they are all separate from one another.

Thus we have the use of variables called I and J throughout each program unit and they are all distinct from one another and altering one of them has no effect at all on the others. This ability to hide information is a very powerful tool, and without it we would find it very difficult to construct large secure bug free programs.

23.3.6 Note 7 – Flexible Design

The QuickSort recursive routine can be replaced with another sorting algorithm and we can maintain the interface to Sort_Data. We can thus decouple the implementation of the actual sorting routine from the defined interface. We would only need to recompile the Sort_Data routine and we could relink using the already compiled main routine, read data routine and print data routines.

23.4 Example 4: Rank two and higher arrays as parameters

23.4.1 Assumed Shape Arrays

When we introduced subroutines we only used scalar parameters. The three preceding examples introduced parameters that were rank one arrays and used explicit shaped dummy arrays. Explicit shape dummy arrays have their shape explicitly specified in a procedure, or just their rank specified, the bounds depending on the value of one or more dummy arguments. Either way this is not very flexible, especially for rank two and higher arrays where extra arguments may be needed to specify the bounds of the dummy arrays.

A much more flexible approach is to declare a dummy array so that it assumes the shape of the actual array argument. Such arrays are called assumed shape arrays. Consider the following program and subroutine segments:

```
PROGRAM Matrix_works
   IMPLICIT NONE
   REAL, DIMENSION (1:50,1:50)::One,Two,Three,One_T
   INTEGER :: I,N
   INTERFACE
      SUBROUTINE Matrix_bits(A,B,C,A_T,N)
         IMPLICIT NONE
         INTEGER, INTENT(IN) :: N
         REAL, DIMENSION (:,:), INTENT(IN) :: A,B
         REAL, DIMENSION (:,:), INTENT(OUT) :: C,A_T
      END SUBROUTINE Matrix_bits
   END INTERFACE
   PRINT *,'Input size of matrices'
   READ*,N
   DO I=1,N
      PRINT*, 'Input row ', I,' of One'
      READ*,One(I,1:N)
   END DO
   DO I=1,N
      PRINT*, 'Input row ', I,' of Two'
      READ*,Two(I,1:N)
   END DO
   CALL Matrix_bits(One,Two,Three,One_T,N)
   PRINT*,' Matrix Three:
   DO I=1,N
      PRINT *,Three(I,1:N)
   END DO
END PROGRAM Matrix_works

SUBROUTINE Matrix_bits(A,B,C,A_T,N)
   IMPLICIT NONE
   INTEGER, INTENT(IN) :: N
   REAL, DIMENSION (:,:), INTENT(IN) :: A,B
   REAL, DIMENSION (:,:), INTENT(OUT) :: C,A_T
   INTEGER ::I
      C=MATMUL(A,B)
      A_T=TRANSPOSE(A)
END SUBROUTINE Matrix_bits
```

In this example in the subroutine Matrix_bits arrays A,B,C and A_T have deferred shape array specifications. When the subroutine is called by the main program the shape of the dummy array arguments will be assumed from their corresponding actual array arguments, One, Two, Three and One_T.

There are several rules and restrictions when using assumed shape arrays:–

- The rank is equal to the number of colons, in this case two;
- The lower bounds of the assumed-shape array are the specified lower bounds, if present, and 1 otherwise. In the example above they are 1 because we haven't specified any lower bounds;
- The upper bounds will be determined on entry to the procedure and will be whatever values are needed to make sure that the extents along each dimension of the dummy argument are the same as those of the actual argument. In his case the upper bounds will be 50 in each dimension;
- An assumed-shape array must not be defined with the POINTER or ALLOCATABLE attribute;
- When using an assumed shape array an interface block is mandatory.

23.5 Summary

We now have a lot of the tools to start tackling problems in a structured and modular way, breaking problems down into manageable chunks and designing subprograms for each of the tasks.

23.6 Problems

1. Here is a program that will create a data file comprising random numbers. Run the program to generate a data file with at least 20,000 real data values. We will use this file as input to the second example.

```
PROGRAM Random
IMPLICIT NONE
INTEGER :: I
REAL :: X
   CALL RANDOM_SEED
   OPEN(UNIT=1,FILE='RANDOM.DAT')
   DO I=1,32767
      CALL RANDOM_NUMBER(X)
      X=X*100
      WRITE (1,FMT='(F6.2)') X
   END DO
END PROGRAM Random
```

2. Run the Quick_Sort program in this chapter with the data file as input. Obtain timing details.

Subroutines: 2

What percentage of the time does the program spend in each subroutine? Is it worth trying to make the sort much more efficient given these timings?

3. Find out if there is a subroutine library like the NAG library available. If there is replace the Quick_Sort recursive subroutine with a suitable routine from that library. What times do you obtain?

4. Try using the operating system SORT command to sort the file. What timing figures do you get now?

Was it worth writing a program?

23.7 Bibliography

So far we have been able to work with a relatively natural language description of the rules and syntax of Fortran. The problems in this and subsequent chapters now require a much more tight and unambiguous syntax definition. One source of this is obviously the standard itself. This is for most people not an option through cost. Given that the standard is itself some 400 pages it is not possible or desirable to duplicate the whole of the standard in this work, copyright permitting (which it is not).

ISO/IEC 1539:1991 (E), *Fortran 90*, 1991

ISO/IEC, *Technical Corrigendum 1 to ISO/IEC 1539:1991*, 1994.

ISO/IEC, *Technical Corrigendum 2 to ISO/IEC 1539:1991*, 1994.

The above comprise the definition of the language. They are aimed more at language lawyers and compiler implementors, rather than the end user. They are in the end the definitive statement regarding the language.

We thus have to look for alternate sources of this information. The best compiler implementations will provide both user documentation and reference documentation. You are strongly urged to see what is supplied with the compiler you use.

Digital Equipment Corporation, *DEC Fortran 90: Language Reference Manual*, Digital Equipment Corporation.

- Very comprehensive and well written coverage of the language. Obviously also contains much implementation specific information. If you use this version of the compiler then an essential companion.

There are a number of books on Fortran that are more reference orientated than user orientated. The following are two of them.

Adams J.C., Brainerd W., S., Martin J.T., Smith B.T., Wagener J.L., *Fortran 90 Handbook: Complete ANSI/ISO Reference*, McGraw Hill.

- Highly recommended. If you don't have access to something like the quality of the above DEC documentation then this is a very useful substitute. Several of the authors were on the X3J3 committee, and this shows in the authority of the contents.

Gehrke W., *Fortran 90 Reference Handbook*, Springer Verlag.

- Comprehensive coverage of the standard. Very reasonably priced source of information.

NAG, *NAG Fortran 77 Library Mark 16*, NAG Ltd.

- A library of over 1,000 numerical and statistical routines written by numerical analysts and full tested and well documented.

NAG, NAG *fl90*, Release 1, NAG Ltd.

- This is a complete revamp of the Fortran 77 library using many of the new features of Fortran 90. It is currently a subset of the Fortran 77 library. There are 141 documented procedures grouped into 44 modules.

24
An Introduction to Modules

Common sense is the best distributed commodity in the world, for every man is convinced that he is well supplied with it.

Descartes.

Aims

The aims of this chapter are to look at the facilities provided in Fortran provided by modules, in particular:–

- the use of a module to aid in the consistent definition of precision throughout a program and sub-programs;
- the use of modules for global data;
- the use of modules for derived data types;
- the use of modules for explicit procedure interfaces;
- the use of modules to package procedures;
- a complete numerical example solving systems of linear equations using Gaussian Elimination;

24 An Introduction to Modules

As summarised in the previous chapter we now have the tools to solve many problems just using a main program and one or more external and internal subprograms. Both external and internal subprograms communicate through their argument lists, whilst internal subprograms have access to data in their host program units.

We now introduce another type of program unit, the module, which is probably one of the most important features of Fortran 90. The purpose of modules is quite different from subprograms. In their simplest form they exist so that anything required by more than one program unit maybe packaged in a module and made available where needed.

The form of a module is:

```
MODULE module_name
...
END MODULE module_name
```

and the information contained within it is made available in the program units that need to access it by the following:–

```
USE module_name
```

The USE statement must be the first statement after the PROGRAM or SUBROUTINE or FUNCTION statement.

In this chapter we will look at:–

- modules for global data;
- modules for derived types;
- modules for explicit interfaces;
- modules containing procedures;

Modules are another program unit and exist so that anything required by more than one program unit maybe packaged in a module and made available where needed.

24.1 Modules for global data

So far the only way that a program unit can communicate with a procedure is through the argument list. Sometimes this is very cumbersome especially if a number of procedures want access to the same data and it means long argument lists. The problem can be solved using modules. For example defining the precision to which you wish to work and any constants defined to that precision which maybe needed by a number of procedures.

An Introduction to Modules

24.1.1 Example 1: Modules for Precision Specification and Constant Definition

In the following example we use a module to define a parameter Long to specify the precision to which we wish to work, and a value for the parameter π. Note that the parameter π is defined to this working precision. We then import the module defining these parameters into the program units that need them.

```
MODULE Precision_defs
   IMPLICIT NONE
   INTEGER, PARAMETER:: Long=SELECTED_REAL_KIND(15,307)
   REAL,PARAMETER::Pi=3.14159265358979_Long
END MODULE Precision_defs

PROGRAM Main
   USE Precision_defs
   IMPLICIT NONE
   REAL(Long)::R,A,C
   INTEGER ::I
   DO I=1,10
      PRINT*,'Radius?'
      READ*,R
      CALL Sub1(R,A,C)
      PRINT *,'For radius ',R, 'Area = ',A,'Circumference = ',C
   END DO
END PROGRAM Main

SUBROUTINE Sub1(Radius,Area,Circum)
   USE Precision_defs
   IMPLICIT NONE
   REAL(Long),INTENT(IN)::Radius
   REAL(Long),INTENT(OUT)::Area,Circum
   Area=Pi*Radius*Radius
   Circum=2.0_Long*Pi*Radius
END SUBROUTINE Sub1
```

24.1.2 Note

In this example we wish to work with the precision specified by the kind type parameter Long in the module Precision_defs. In order to do this we use the statement

```
USE Precision_defs
```

inside the program unit before any declarations. The kind type parameter Long is then used with all the REAL type declaration e.g.

```
REAL (Long):: R ,A,C
```

To make sure that all floating point calculations are performed to the working precision specified by Long any constants such as 2.0 in subroutine Sub1 are specified as const_Long, e.g.

```
2.0_Long
```

Note also that we define things once and use them on two occasions, i.e. we define the precision once and use this definition in both the main program and the subroutine.

24.1.3   Example 2: Constant Definition and Array Definition

The following example uses one module containing a number of constants and a second module containing an array definition. We have again the use of modules to define once and used on two or more occasions.

```
MODULE Consts
    IMPLICIT NONE
    REAL,PARAMETER:: A=15.0,B=22.7,C=3.691
END MODULE Consts

MODULE Data
    IMPLICIT NONE
    INTEGER,PARAMETER::No_of_Readings=100
    REAL,DIMENSION(1:No_of_readings)::X,Y
END MODULE Data

SUBROUTINE Stat(Arg1,Arg2,Arg3)
    USE Data
    IMPLICIT NONE
    REAL,INTENT(OUT)::Arg1,Arg2,Arg3
    INTEGER::I
    Arg1=0.0; Arg2=0.0; Arg3=0.0
    DO I=1,No_of_readings
       Arg1 = Arg1+X(I)*X(I)
       Arg2 = Arg2+Y(I)*Y(I)
       Arg3 = Arg3+X(I)*Y(I)
    END DO
    Arg3=Arg3*Arg3
END SUBROUTINE Stat

PROGRAM Main
USE Consts
USE Data
```

```
IMPLICIT NONE
INTERFACE
   SUBROUTINE Stat(Arg1,Arg2,Arg3)
      USE Data
      IMPLICIT NONE
      REAL,INTENT(OUT)::Arg1,Arg2,Arg3
   END SUBROUTINE Stat
END INTERFACE
   CHARACTER (LEN=40):: File_name
   INTEGER:: I
   REAL:: Sum,SUM_x_2,Sum_y_2,Sum_xy_2
!
   PRINT *,'Data file name?'
   READ(*,'(A)')File_name
   OPEN(UNIT=10,FILE=File_name,STATUS='OLD')
   DO I=1,No_of_readings
      READ(10,*)X(I),Y(I)
   END DO
   CALL Stat(Sum_x_2,Sum_y_2,Sum_xy_2)
   Sum= (A*Sum_x_2 + B*Sum_y_2) / (C*Sum_xy_2)
   PRINT *,'Sum = ',Sum
!
END PROGRAM Main
```

24.2 Modules for derived data types

When using derived data types and passing them as arguments to subroutines, both the actual arguments and dummy arguments must be of the same type, that is they must be declared with reference to the same type definition. The only way this can be achieved is by using modules. The user defined type is declared in a module and each program unit that requires that type **use**s the module.

24.2.1 Example 3: Person Data Type

In this example we have a user defined type Person which we wish to use in the main program and pass arguments of this type to the subroutines Read_data and Stats. In order to have the type Person available to two subroutines and the main program we have defined Person in a module Personal_details and then made the module available to each program unit with the statement:-

```
USE Personal_details
```

We also have the use of interface block to provide the ability to develop the overall solution in stages.

```
MODULE Personal_details
   IMPLICIT NONE
   TYPE Person
      REAL:: Weight
      INTEGER :: Age
      CHARACTER :: Sex
   END TYPE Person
END MODULE Personal_details

PROGRAM Survey
   USE Personal_details
   IMPLICIT NONE
   INTEGER ,PARAMETER:: Max_no=100
   TYPE (Person), DIMENSION(1:Max_no) :: Patient
   INTEGER :: I,No_of_patients
   REAL :: Male_average, Female_average
INTERFACE

   SUBROUTINE Read_data(Data,Max_no,No)
      USE Personal_details
      IMPLICIT NONE
      TYPE (Person), DIMENSION (:), INTENT(OUT):: Data
      INTEGER, INTENT(OUT):: No
      INTEGER, INTENT(IN):: Max_no
   END SUBROUTINE Read_Data

   SUBROUTINE Stats(Data,No,M_a,F_a)
      USE Personal_details
      IMPLICIT NONE
      TYPE(Person), DIMENSION (:) :: Data
      REAL:: M_a,F_a
      INTEGER :: No
   END SUBROUTINE Stats

END INTERFACE
!
   CALL Read_data(Patient,Max_no,No_of_patients)
   CALL Stats(Patient,No_of_patients,Male_average,Female_average)
   PRINT*, 'Average male weight is ',Male_average
   PRINT*, 'Average female weight is ',Female_average
END PROGRAM Survey
```

An Introduction to Modules

```fortran
SUBROUTINE Read_Data(Data,Max_no,No)
   USE Personal_details
   IMPLICIT NONE
   TYPE (PERSON), DIMENSION (:), INTENT(OUT)::Data
   INTEGER, INTENT(OUT):: No
   INTEGER, INTENT(IN):: Max_no
   INTEGER :: I
   DO
      PRINT *,'Input number of patients'
      READ *,No
      IF ( No > 0 .AND. No <= Max_no) EXIT
   END DO
   DO I=1,No
      PRINT *,'For person ',I
      PRINT *,'Weight ?'
      READ*,Data(I)%Weight
      PRINT*,'Age ?'
      READ*,Data(I)%Age
      PRINT*,'Sex ?'
      READ*,Data(I)%Sex
   END DO
END SUBROUTINE Read_Data

SUBROUTINE Stats(Data,No,M_a,F_a)
   USE Personal_details
   IMPLICIT NONE
   TYPE(Person), DIMENSION(:)::Data
   REAL :: M_a,F_a
   INTEGER:: No
   INTEGER :: I,No_f,No_m
   M_a=0.0; F_a=0.0;No_f=0; No_m =0
   DO I=1,No
      IF(Data(I)%Sex == 'M' .OR. Data(I)%Sex == 'm') THEN
         M_a=M_a+Data(I)%Weight
         No_m=No_m+1
      ELSEIF(Data(I)%Sex == 'F'.OR. Data(I)%Sex == 'f') THEN
         F_a=F_a +Data(I)%Weight
         No_f=No_f+1
      ENDIF
   END DO
   M_a = M_a/No_m ; F_a = F_a/No_f
END SUBROUTINE Stats
```

24.3 Modules for explicit procedures interfaces

When we introduced subroutines we also introduced the interface block, and have seen how useful it is as an extra security measure at compile time, to check that actual and dummy arguments agree. Interface blocks are mandatory under the following conditions:–

- for argument types like pointers, assumed shape arrays and assumed shape dummy arguments;
- when defining generic procedures – see chapter 26;
- when operator overloading – see chapter 26;
- when defining new meaning to the assignment symbol;

When programs start to become larger with many procedure references it becomes tedious and time consuming to keep repeating interface blocks. One way round this is to place one or more procedure interfaces in a module and then use the module in each program unit as required. This is also strongly recommended for programs where we use the same interface in two or more places. The golden rule is define once and use many times.

24.3.1 Example: Using QuickSort

Using the Quicksort example in the previous chapter we have:–

```
MODULE Read_Interface
INTERFACE
   SUBROUTINE Read_data ( File_name,Raw_Data,How_Many)
      IMPLICIT NONE
      CHARACTER (LEN=*),INTENT(IN)     :: File_Name
      INTEGER, INTENT(IN) :: How_Many
      REAL, INTENT (OUT), DIMENSION (How_Many) :: Raw_Data
   END SUBROUTINE Read_Data
END INTERFACE
END MODULE Read_Interface

MODULE Sort_Interface
INTERFACE
   SUBROUTINE Sort_Data(Raw_Data, How_Many)
      IMPLICIT NONE
      INTEGER , INTENT(IN) :: How_Many
      REAL , INTENT(IN), DIMENSION (How_Many) :: Raw_Data
   END SUBROUTINE Sort_Data
END INTERFACE
END MODULE Sort_Interface
```

```
MODULE Print_Interface
INTERFACE
   SUBROUTINE Print_Data (Raw_Data,How_Many)
      IMPLICIT NONE
      INTEGER , INTENT(IN)     :: How_Many
      REAL , INTENT(IN), DIMENSION (How_Many) :: Raw_Data
   END SUBROUTINE Print_Data
END INTERFACE
END MODULE Print_Interface
```

and then the main program would look like:–

```
PROGRAM Sort_Example01
USE Read_Interface
USE Sort_Interface
USE Print_Interface
IMPLICIT NONE
INTEGER :: How_many
   ...
   CALL Read_Data(File_Name, Raw_Data, How_Many)
   ...
END PROGRAM Sort_Example01
```

24.4 Modules containing procedures

A module can also contain one or more procedures, and takes the form:–

```
MODULE Procs
...
CONTAINS
   ...
   SUBROUTINE sub
   ...
   END SUBROUTINE sub
   ...
END MODULE Procs
```

This is a very powerful feature. The procedures inside the module are called **module procedures**. Like internal procedures they have access to any data and type definitions declared in the module, and their interfaces are explicit to each other.

When designing a suite of subprograms, related subroutines and functions can be packed into one module.

For example if you are designing a suite of sub-programs to solve ordinary differential equations you may want to offer three methods:-

- Runge Kutta
- Adams Bashforth
- Gear

The subroutines implementing these three algorithms could be packaged into one module:-

```
MODULE Odes
IMPLICIT NONE
INTEGER, PARAMETER:: Long=SELECTED_REAL_KIND(15,307)
REAL (Long) :: Step_interval = 100.0_Long
REAL (Long) :: Smallest_step = 1.0E-6_Long
...
CONTAINS
   SUBROUTINE RK(...)
   .
   END SUBROUTINE RK
   SUBROUTINE AB(...)
   .
   END SUBROUTINE AB
   SUBROUTINE Gear(...)
   .
   END SUBROUTINE Gear
END MODULE Odes
```

All the subroutines in the module have access to the Kind type parameter Long and the two variables Step_interval and Smallest_step.

A main program to call any of these subroutines could be of the form:-

```
PROGRAM Diff_Eqns
USE Odes
IMPLICIT NONE
REAL (Long), DIMENSION (:), ALLOCATABLE :: Y,F
REAL (Long) ::T,A,B
CHARACTER (LEN=2) :: Method
   PRINT*,'Method? (RK, AB or G)'
   READ(*,'(A)')Method
   SELECT CASE (Method)
      CASE ('RK','rk','Rk','rK')
         CALL RK(...)
```

```
           ...
     CASE ('AB','ab', 'Ab','aB')
        CALL AB(...)
        ...
     CASE ('G','g')
        CALL Gear(...)
        ....
     CASE DEFAULT
        PRINT*,'No valid method chosen'
   END SELECT
END PROGRAM Diff_Eqns
```

This is how Fortran 90 libraries will be designed in the future. The Numerical Algorithms Group (NAG) have been the providers of a Fortran numerical and statistical library for a number of years. They have recently released a completely redesigned Fortran 90 library, albeit a subset of the popular Fortran 77 library. The Fortran 90 library, fl90, is organised into chapters and within each chapter there are one or more modules, each module containing one of more procedures, together with type definitions, global data etc.

For instance in release 1 chapter 3 - *Special Functions* has seven modules. One such module is called nag_gamma_fun which contains four procedures:–

Procedure Name	Action
nag_gamma	Gamma function
nag_log_gamma	Log gamma function
nag_polygamma	Polygamma function
nag_incompl_gamma	Incomplete gamma functions

24.5 Example 4 – The Solution of Linear Equations Using Gaussian Elimination

At this stage we have introduced many of the concepts needed to write numerical code, and so we have included a popular algorithm, Gaussian Elimination together with a main program which uses it, and a module, to brings together many of the features covered so far.

Finding the solution of a system of linear equations is very common in scientific and engineering problems, either as a direct physical problem or indirectly for example as the result of using finite difference methods to solve a partial differential equation. We will restrict ourselves to the case where the number of equations and the number of unknowns are the same. The problem can be defined as:–

$$a_{11}x_1 + a_{12}x_2 \ldots + a_{1n}x_n = b_1$$
$$a_{21}x_2 + a_{22}x_2 \ldots + a_{n2}x_n = b_2$$
$$\vdots$$
$$a_{n1}x_1 + a_{n2}x_n \cdots + a_{nn}x_n = b_n$$

(1)

or

$$\begin{pmatrix} a_{11} & a_{12} & \ldots & a_{1n} \\ a_{21} & a_{22} & \ldots & a_{2n} \\ \vdots & \vdots & \ldots & \vdots \\ a_{n1} & a_{n2} & \ldots & a_{nn} \end{pmatrix} \begin{pmatrix} x_1 \\ x_2 \\ \vdots \\ x_n \end{pmatrix} = \begin{pmatrix} b_1 \\ b_2 \\ \vdots \\ b_n \end{pmatrix}$$

which can be written as:

$$A x = b \qquad (2)$$

where A is the n x n coefficient matrix, b is the right hand side vector and x is the vector of unknowns. We will also restrict ourselves to the case where A is a general real matrix.

Note, there is a unique solution to (2) if the inverse, A^{-1}, of the coefficient matrix A, exists. However the system should never be solved by finding A^{-1} and then solving $A^{-1} b = x$ because of the problems of rounding error and the computational costs.

A well known method for solving (2) is Gaussian Elimination, where multiples of equations are subtracted from others so that the coefficients below the diagonal become zero, producing a system of the form: –

$$\begin{pmatrix} a_{11}^* & a_{12}^* & a_{13}^* & \ldots & a_{1n}^* \\ 0 & a_{22}^* & a_{23}^* & \ldots & a_{2n}^* \\ 0 & 0 & a_{33}^* & \ldots & a_{3n}^* \\ \vdots & \vdots & \vdots & \ldots & \vdots \\ 0 & 0 & 0 & 0 & a_{nn}^* \end{pmatrix} \begin{pmatrix} x_1 \\ x_2 \\ x_3 \\ \vdots \\ x_n \end{pmatrix} = \begin{pmatrix} b_1^* \\ b_2^* \\ b_3^* \\ \vdots \\ b_n^* \end{pmatrix}$$

where A has been transformed into an upper triangular matrix. By a process of *backward substitution* the values of x *drop* out.

The subroutine Gaussian_Elimination implements the Gaussian elimination algorithm with *partial pivoting* which makes sure that the multipliers are less than 1 in magnitude, by interchanging rows if necessary. This is to try and prevent the build up of errors.

This implementation is based on two LINPACK routines SGEFA and SGESL and a Fortran 77 subroutine written by Tim Hopkins and Chris Phillips in their excellent book *Numerical Methods in Practice*.

An Introduction to Modules

The matrix A and vector B are passed to the subroutine Gaussian_Elimination and on exit both A and B are overwritten. Mathematically Gaussian elimination is described as working on rows, and using partial pivoting row interchanges may be necessary. Due to Fortran's row element ordering, to implement this algorithm efficiently it works on columns rather than rows by interchanging elements within a column if necessary.

```
PROGRAM Solve
   USE Precisions
   IMPLICIT NONE
   INTEGER :: I,N
   REAL (Long), ALLOCATABLE:: A(:,:),B(:),X(:)
   LOGICAL:: Singular
INTERFACE
   SUBROUTINE Gaussian_Elimination(A,N,B,X,Singular)
      USE Precisions
      IMPLICIT NONE
      INTEGER, INTENT(IN)::N
      REAL (Long), INTENT (INOUT) :: A(:,:),B(:)
      REAL (Long), INTENT(OUT)::X(:)
      LOGICAL, INTENT(OUT) :: Singular
   END SUBROUTINE Gaussian_Elimination
END INTERFACE
!
   PRINT *,'Number of equations?'
   READ *,N
   ALLOCATE(A(1:N,1:N),B(1:N),X(1:N))
   DO I=1,N
      PRINT *,'Input elements of row ',I,' of A'
      READ*,A(I,1:N)
      PRINT*,'Input element ',I,' of B'
      READ *,B(I)
   END DO
   CALL Gaussian_Elimination(A,N,B,X,Singular)
   IF(Singular) THEN
      PRINT*, 'Matrix is singular'
   ELSE
      PRINT*, 'Solution X:'
      PRINT*,X(1:N)
   ENDIF
END PROGRAM Solve
```

```fortran
SUBROUTINE Gaussian_Elimination(A,N,B,X,Singular)
! Routine to solve a system Ax=b using Gaussian Elimination
! with partial pivoting
! The code is based on the Linpack routines SGEFA and SGESL
! and operates on columns rather than rows!
   USE Precisions
   IMPLICIT NONE
! Matrix A and vector B are over-written
! Arguments
   INTEGER, INTENT(IN):: N
   REAL (Long),INTENT(INOUT):: A(:,:),B(:)
   REAL (Long),INTENT(OUT)::X(:)
   LOGICAL,INTENT(OUT)::Singular
! Local variables
   INTEGER::I,J,K,Pivot_row
   REAL (Long):: Pivot,Multiplier,Sum,Element
   REAL (Long),PARAMETER::Eps=1.E-13_Long
!
! Work through the matrix column by column
!
   DO K=1,N-1
!
!  Find largest element in column K for pivot
!
   Pivot_row=MAXVAL( MAXLOC( ABS( A(K:N,K) ) ) ) + K - 1
!
! Test to see if A is singular - if so return to main program
!
      IF(ABS(A(Pivot_row,K)) <= Eps) THEN
         Singular=.TRUE.
         RETURN
      ELSE
         Singular = .FALSE.
      ENDIF
!
! Exchange elements in column K if largest is
! not on the diagonal
!
      IF(Pivot_row /= K) THEN
         Element=A(Pivot_row,K)
         A(Pivot_Row,K)=A(K,K)
         A(K,K)=Element
```

```
            Element=B(Pivot_row)
            B(Pivot_row)=B(K)
            B(K)=Element
         ENDIF
!
! Compute multipliers - elements of column K below diagonal
! are set to these multipliers for use in elimination later on
!
      A(K+1:N,K) = A(K+1:N,K)/A(K,K)
!
! Row elimination performed by columns for efficiency
!
      DO J=K+1,N
         Pivot = A(Pivot_row,J)
         IF(Pivot_row /= K) THEN   ! Swap if pivot row is not K
            A(Pivot_row,J)=A(K,J)
            A(K,J)=Pivot
         ENDIF
         A(K+1:N,J)=A(K+1:N,J)-Pivot* A(K+1:N,K)
      END DO
!
! Apply same operations to B
!
      B(K+1:N)=B(K+1:N)-A(K+1:N,K)*B(K)
   END DO
!
! Backward substitution
!
   DO I=N,1,-1
      Sum = 0.0
      DO J= I+1,N
         Sum=Sum+A(I,J)*X(J)
      END DO
      X(I)=(B(I)-Sum)/A(I,I)
   END DO
END SUBROUTINE Gaussian_Elimination

MODULE Precisions
INTEGER,PARAMETER:: Long=SELECTED_REAL_KIND(15,307)
END MODULE Precisions
```

24.5.1 Notes

1. A module, Precisions, has been used to define a kind type parameter, Long, to specify the floating point precision to which we wish to work. This module is then used by the main program and the subroutine, and the kind type parameter Long is used with all the REAL type definitions and with any constants, e.g.

REAL(Long), PARAMETER :: Eps=1.E-13_Long

2. In the main program Matrix A and vectors B and X are declared as deferred shape arrays, by specifying their rank only and using the ALLOCATABLE attribute. Their shape is determined at run time when the variable N is read in and then the statement:–

ALLOCATE(A(1:N,1:N), B(1:N), X(1:N))

is used.

3. Using intrinsic functions MAXVAL and MAXLOC

In the context of subroutine Gaussian_Elimination we have used:-

MAXVAL (MAXLOC (ABS (A (K:N,K)))) + K - 1

Breaking this down:-

MAXLOC (ABS (A (K:N,K)))

takes the rank one array

$(\,|A(K,K)|\,,|A(K+1,K)|\,,\ldots|A(N,K)|\,)$ \hfill (1)

where $|A(K,K)| = $ ABS(A(K,K)) and of length N- K + 1. It returns the position of the largest element as a rank one array of size one, e.g. (L)

Applying MAXVAL to this rank one array (L) returns L as a scalar. L being the position of the largest element of array (1).

What we actually want is the position of the largest element of (1), but in the Kth column of matrix A. We therefore have to add K-1 to L to give the actual position in column K of A.

24.6 Notes on Module Usage and Compilation

If we only have one file comprising all of the program units (main program, modules, functions and subroutines) then there is little to worry about. However it is recommended that larger scale programs are developed as a collection of files with related program units in each file, or even one program unit per file. This is more productive in the longer term. This will however lead to problems with modules unless we compile each module **before** we use it in other program units.

An Introduction to Modules

Secondly we must use one directory or sub-directory so that the compiler and linker can find each program unit.

Thirdly we must be aware of the file naming conventions used by each compiler implementation we work with. Consider the following:–

Fortran Module Name	Salford/Nag under DOS	NAG OPEN VMS	DEC F90 OPEN VMS
Precisions	MOD_4689.MOD	PRECISIONS.MOD	PRECISIONS.F90$MOD
Node_Type_Def	MOD_4024.MOD	NODE_TYPE_DEF.MOD	NODE_TYPE_DEF.F90$MOD

On the PC under DOS this can be a problem as with an eight character name and three character extension we cannot possibly map valid Fortran 90 module names onto eleven characters.

On the VAX under OPENVMS and with the NAG Fortran compiler we do not have any problems with naming conventions. The complete file name has a .MOD extension.

On the DEC Alpha under OPENVMS with the DEC Fortran 90 compiler the complete file name has a .F90$MOD extension, following in the DEC VMS tradition of having $ in system names.

24.7 Summary

We have now introduced the concept of a module, another type of program unit, probably one of of the most important features of Fortran 90. We have seen in this chapter how they can be used

- define global data
- define derived data types
- contain explicit procedure interfaces
- package together procedures

This is a very powerful addition to the language especially when constructing large programs and procedure libraries.

24.8 Problems

1. Write two functions, one to calculate the volume of a cylinder $\pi r^2 l$ where the radius is r and the length is l, and the other to calculate the area of the base of the cylinder πr^2. Define π as a

parameter in a module which is used by the two functions. Now write a main program which prompts the user for the values of r and l, calls the two functions and prints out the results.

24.9 Bibliography

Dongarra, J., Bunch, J.R., Moler, C.B., and Stewart, G.W. *LINPACK User's Guide*. SIAM Publications, Philadelphia, 1979.

- This Fortran 77 package is for the solution of simultaneous systems of linear algebraic equations. Special subroutines are included for many common types of coefficient matrices. The source is available through NETLIB. See chapter 28 for more details.

Hopkins T., Phillips C., *Numerical Methods in Practice, using the NAG library*. Addison Wesley.

- This is a very good practical introduction to numerical analysis, with the aim of guiding users to the more commonly used routines in the NAG Fortran 77 library. It does this by introducing topics, giving some background, advantages and disadvantages, and the Fortran 77 code for some of the more well known algorithms. It then introduces the appropriate NAG routine with a brief discussion of its use, calling sequence and any error reporting facilities.
  We've found this invaluable for many of our students who are users of the NAG library but not well versed with numerical analysis. Maybe we will see a Fortran 90 version of this book in the near future?

NAG, NAG *fl90*, Release 1.

- This library is a complete revamp of their Fortran 77 library. It is a subset currently. There are 141 documented procedures grouped into 44 modules.

25
Formal Syntax and Some Additional Features

Once upon a time, a very long time ago now, about last Friday, Winnie-the-Pooh lived in a forest all by himself under the name of Sanders.

(*"What does 'under the name' mean"* asked Christopher Robin, *"It means he had the name over the door in gold letters and lived under it"*...)

A. A. Milne., *Winnie-the-Pooh.*

Aims

The aims of this chapter are to introduce formal definitions of a number of concepts and also look at some of the more advanced features of Fortran 90, in particular:–

- program units
- functions
- subroutines
- modules
- scope
- modules and scope
  - public
  - private
  - rename
- optional arguments

25 Formal Syntax and Some Additional Features

In the coverage so far we have relied on quite lax definitions of the concepts introduced choosing to provide concrete examples, rather than overwhelm the reader with rather indigestible chunks of grammar and syntax. In this chapter we look more formally at some of the earlier concepts as the reliable construction of larger programs requires a more thorough understanding of these concepts.

We will also look at a number of advanced features of Fortran 90.

This material draws on chapter two of the standard (Fortran terms and concepts) and Annex D of the standard.

25.1 Program Units

Program units are the fundamental building block of Fortran 90. A program unit may be one of:–

- a main program
- a function
- a subroutine
- a module

and for the first three there will be a specification part, an execution part. A module does not have an executable component, unless there is a **contain**ed subprogram.

25.2 Procedure – Function or Subroutine

A procedure is either a function or subroutine. A function is a procedure that is invoked in an expression, whilst a subroutine is a procedure that is invoked by the CALL statement.

25.2.1 Internal Procedure

An internal procedure is one that is CONTAINed within a program unit. An internal procedure is local to its host. There are examples of this with the Stirling function in chapter 14 and the quick sort recursive subroutine in chapter 23.

25.3 Module

The module program unit is a very powerful feature of Fortran 90. They are used to:–

- provide access to type definitions;
- provide access to global data;
- provide user defined operators;
- provide the ability to manage data types and valid operations on those types – data abstraction;

- provide packaging of functions and subroutines;

Access to some or all of the components of modules can be controlled by the use of a number of additional extensions that we will cover later, namely PRIVATE, PUBLIC and RENAME.

25.4 Executable Statements

Program execution is a sequence of computational actions.

25.5 Statement Ordering

The following figure summarises the statement ordering.

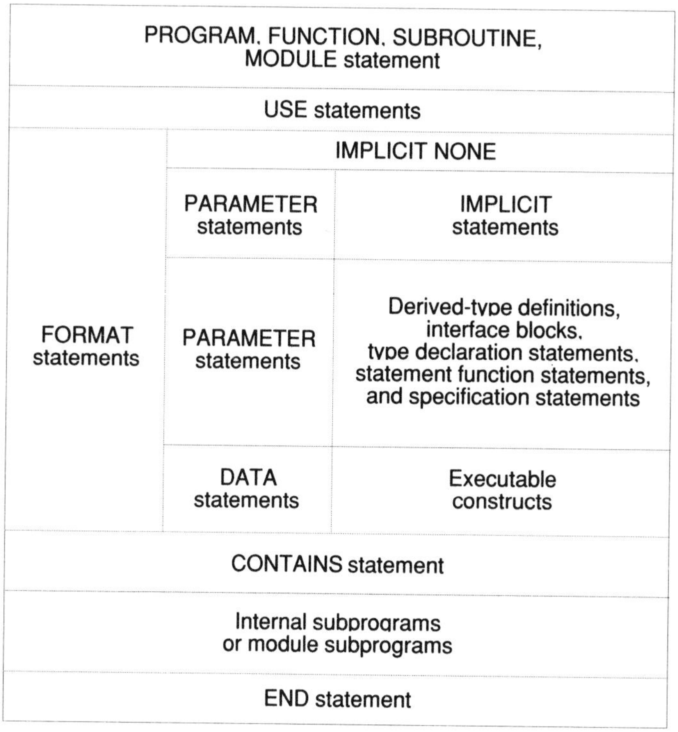

PROGRAM, FUNCTION, SUBROUTINE, MODULE statement		
USE statements		
	IMPLICIT NONE	
	PARAMETER statements	IMPLICIT statements
FORMAT statements	PARAMETER statements	Derived-type definitions, interface blocks, type declaration statements, statement function statements, and specification statements
	DATA statements	Executable constructs
CONTAINS statement		
Internal subprograms or module subprograms		
END statement		

Figure 2

FORMAT statements may appear anywhere between the USE statement and the CONTAINS statement.

The following table summarises the usage of the various statements within individual scoping units.

Kind of Scoping unit	Main program	Module	External sub program	Module sub program	Internal sub program	Interface body
USE	Y	Y	Y	Y	Y	Y
FORMAT	Y	N	Y	Y	Y	N
Misc Dec [1]	Y	Y	Y	Y	Y	Y
Derived type definition	Y	Y	Y	Y	Y	Y
Interface block	Y	Y	Y	Y	Y	Y
Executable statement	Y	N	Y	Y	Y	N
CONTAINS	Y	Y	Y	Y	N	N

[1] Misc Dec (Miscellaneous declaration) are PARAMETER statements, IMPLICIT statements, type declaration statements and specification statements.

25.6 Entities

An entity may be identified by:–

- a name
- a label
- external i/o unit number
- an operator symbol
- an assignment symbol

and by means of association an entity may be referred to by the same identifier or a different identifier in a different scoping unit, or by a different identifier in the same scoping unit.

25.7 Scope and Association

In the simplest case scope refers to the places within a program where a name has meaning. Where we only have one program unit then we do not have any problems with the concept of scope as all named entities are available within that program unit.

The problem becomes more complex when we start constructing programs that comprise one or more procedures and use internal procedures and external libraries of routines.

Formal Syntax and Some Additional Features

A program consists of one or more non-overlapping scoping units. A scoping unit is:–

- a program unit or subprogram, excluding derived type definitions, procedure interface bodies, and subprograms contained within it.
- a procedure interface body, excluding any derived type definitions and procedure interface bodies contained within it;
- a derived type definition;

and a scoping unit that encloses another scoping unit is called the host scoping unit.

The USE statement provides the means by which a scoping unit accesses named data objects, derived data types, procedures, generic identifiers in a module. The entities in the scoping unit are said to be use associated with the entities in the module.

Using the sorting program in chapter 23 as an example we have:–

- Main program

    Variables: How_Many, File_Name, Raw_Data, Count, Count_Rate, T1, T2, T3, T4, I, J
- SYSTEM_CLOCK(Count,Count_Rate)

    Duumy arguments: Count, Count_Rate

    No accessible variables
- Read_Data(File_Name,Raw_Data,How_Many)

    Dummy arguments: File_Name,Raw_Data,How_Many

    Local variable: I
- Sort_Data(Raw_Data,How_Many)

    Dummy arguments: Raw_Data,How_Many

    Contains:

    Quick_Sort(L,R)

    Dummy arguments: L,R

    Local variables: I,J,V,T
- Print_Data(Raw_Data,How_Many)

    Dummy arguments: Raw_Data,How_Many

    Local variable: I

The variables How_Many, File_Name, Raw_Data, Count and Count_Rate, are declared within the main program. We make them available in the four subroutines by argument association.

The variables T1, T2, T3, T4, I and J are declared within the main program. These variables are purely local to the main program.

The variable I in the Read_Data routine is local to that routine and is not available elsewhere.

The variables Raw_Data and How_Many in the Sort_Data subroutine are globally available to the Quick Sort subroutine as QuickSort is CONTAINed within the Sort Data subroutine. This is called host association.

The variables I, J, V and T are purely local to the QuickSort routine and not available elsewhere. Note also as this routine is recursive there will be multiple instances of this routine in existence as the sort proceeds and therefore multiple and completely separate instances of these four variables.

The Print_Data routine has How_Many and Raw_Data available via argument association and I is purely local.

Thus Fortran provides complete decoupling of entity names between program units automatically. The only way that communication can take place is via argument association and host association in the above example.

25.8 Modules and Scope

Modules provide a very powerful way to make data available throughout program units. In the example at the start of the previous chapter we had a module that contained a definition for π. We then **use**d this module in each subprogram unit that needed it. Thus we have the ability with modules to extend the way in which we make named entities available throughout program units.

The power of modules increases further when we use them to make data types available and operations on these data types.

When a module contains an executable procedure the problem immediately arises as to the accessibility of variables within the module, and potential name conflicts. Fortran provides a number of facilities to control things in a situation like this.

25.8.1 Public and Private Attributes

By default entities within a module are PUBLIC, i.e. they are visible to every program unit in which the module is made available. An entity is one of:–

> type definition
> variable
> nonintrinsic function
> named constants

Entities that are made available via the USE statement are said to be accessed by *use association*.

If we wished to hide some entities in a module we would add an attribute of PRIVATE to their declaration, e.g. in section 24.4 in the previous chapter we could make Step_Interval and Small-

Formal Syntax and Some Additional Features

est_Step invisible to the main program in which the module is **use**d. We could do this in the following way:–

```
MODULE ODES
IMPLICIT NONE
INTEGER , PARAMETER :: Long=SELECTED_REAL_KIND(15,307)
REAL (Long) , PRIVATE :: Step_Interval =100.0_Long
REAL (Long) , PRIVATE :: Smallest_Step =1.0E-6_Long
...
END MODULE ODES
```

Step_Interval and Smallest_Step are now hidden from the main program in which the module is used but the parameter Long is available because by default it has PUBLIC attribute.

There is an additional form of PRIVATE that enables us to hide nonintrinsic subroutines, generic names and operator and assignment overloading. See chapter 26 for an example of operator and assignment overloading.

There are also additional forms of the USE statement that together with the above two forms provide most of the flexibility we require in this area.

25.8.2 USE, ONLY and Rename

In the examples so far we have only had simple forms of the USE statement, e.g.

```
USE Module_Name
```

There are a number of additional forms that are very useful. One provides the ability to import **only** a subset of the entities from the used module, and is of the form:–

```
USE Module-Name , ONLY : access-list
```

where access-list is any of the **public** entities in the module, e.g.

```
USE String_Module , ONLY : X,Y
```

and only the variables X and Y in module String_Module are available to the program unit in which this use statement appears.

The second form provides the ability to rename entities to avoid name conflicts, and is of the form:–

```
USE Module-Name [local-name=>] module-entity-name
```

The local-name (if present) is the new name of that entity in the using program unit, e.g.

```
USE Tree_Module , Tree_Walk => TW
```

The module variable Tree_Walk is known as TW in the program unit in which the use statement appears.

25.9 Keyword and Optional Arguments

The examples of procedures so far have assumed that the dummy arguments and the corresponding arguments are in the same position, i.e. we are using positional arguments. Fortran 90 also provides the ability to supply the actual arguments to a procedure by keyword, and hence in any order.

To do this the name of the dummy argument is referred to as the keyword and is specified in the actual argument list in the form:-

```
dummy-argument = actual-argument
```

To illustrate this, let us consider a subroutine to solve ordinary differential equations. The full subroutine and explanation are given in the next chapter.

```
SUBROUTINE Runge_Kutta_Merson(Y,FUN,IFAIL,N,A,B,TOL)
```

where A is the initial point, B is the end point at which the solution is required, TOL is the accuracy to which the solution is required and N is the number of equations. The rest of the dummy arguments are explained in the next chapter.

The subroutine can be called as follows:-

```
CALL Runge_Kutta_Merson(Y,Fun1,IFAIL,A=0.0,B=8.0,Tol=1.0E-6,N=3)
```

where the dummy arguments A,B,Tol and N are now being used as keywords. The use of keyword arguments make the code easier to read and decreases the need to remember their precise position in the argument list.

Also with Fortran 90 comes the ability to specify that an argument is optional. This is very useful when designing procedures for use by a range of programmers. Inside a procedure defaults can be set for the optional arguments providing an easy-to-use interface, at the same time allowing sophisticated users a comprehensive interface.

To declare a dummy argument to be optional the OPTIONAL attribute can be used. For example the last dummy argument Tol for the subroutine Runge_Kutta_Merson could be declared to be optional (although internally in the subroutine the code would have to be changed to allow for this) e.g.

```
SUBROUTINE Runge_Kutta_Merson(Y,FUN,IFAIL,N,A,B,Tol)
   REAL(Long), INTENT(INOUT), OPTIONAL :: Tol
```

and because it is at the end of the dummy argument list, calling the subroutine with a positional argument list, Tol can be omitted, e.g.

```
CALL Runge_Kutta_Merson(Y,Fun1,IFAIL,N,A,B)
```

Formal Syntax and Some Additional Features

The code of the subroutine will need to be changed to check to see if the argument Tol is supplied, the intrinsic function PRESENT being available for this purpose.

A number of points need to be noted when using keyword and optional arguments:–

- If all the actual arguments use keywords they may appear in any order.

- When only some of the actual arguments use keywords the first part of the list must be positional followed by keyword arguments in any order.

- When using a mixture of positional and keyword arguments, once a keyword argument is used all subsequent arguments must be specified by keyword.

- If an actual argument is omitted the corresponding optional dummy argument must not be redefined or referenced, except as an argument to the PRESENT intrinsic function.

- If an optional dummy argument is at the end of the argument list then it can just be omitted from the actual argument list.

- Keyword arguments are needed when an optional argument not at the end of an argument list is omitted, unless all the remaining arguments are omitted as well.

- Keyword and optional arguments require explicit procedure interfaces, i.e. the procedure must be internal, a module procedure or have an interface block available in the calling program unit.

As we have already seen with some of the intrinsic procedures it is possible to miss out one or more arguments, e.g. SYSTEM_CLOCK.

25.10 Syntax Summary of Some Frequently used Fortran Constructs

The following provides simple syntactical definitions for of some of the more frequently used parts of Fortran 90. Appendix G provides a full BNF description. This is Annex D of the standard.

25.10.1 Main Program

```
PROGRAM [ program-name ]
   [ specification-construct ] ...
   [ executable-construct ] ...
   [CONTAINS
   [ internal procedure ] ... ]
END [ PROGRAM [ program-name ] ]
```

25.10.2 Subprogram

```
procedure heading
   [ specification-construct ] ...
   [ executable-construct ] ...
   [CONTAINS
```

```
    [ internal procedure ] ... ]
procedure ending
```

25.10.3 Module

```
MODULE name
    [ specification-construct ] ...
    [CONTAINS
    subprogram
    [ subprogram ] ... ]
END [ MODULE  [ module-name ]
```

25.10.4 Internal Procedure

```
procedure heading
    [ specification construct ] ...
    [ executable construct ] ...
procedure ending
```

25.10.5 Procedure heading

```
[ RECURSIVE ] [ type specification ] FUNCTION function-name &
    ( [ dummy argument list ] ) [ RESULT ( result name ) ]
[ RECURSIVE ] SUBROUTINE subroutine name &
    [ ( [ dummy argument list ] ) ]
```

25.10.6 Procedure ending

```
END [ FUNCTION [ function name ] ]
END [ SUBROUTINE [ subroutine name ] ]
```

25.10.7 Specification construct

```
derived type definition
interface block
specification statement
```

25.10.8 Derived Type definition

```
TYPE [[ , access specification ] :: ] type name
    [ PRIVATE ]
    [ SEQUENCE ]
    [ type specification [[ , POINTER ] :: ] component specification
list ]
    ...
    END TYPE [ type name ]
```

25.10.9 Interface block

```
INTERFACE [ generic specification ]
    [ procedure heading
      [ specification construct ] ...
    procedure ending ] ...
    [ MODULE PROCEDURE module procedure name list ] ...
END INTERFACE
```

25.10.10 Specification statement

```
ALLOCATABLE [ :: ] allocatable array list
DIMENSION array dimension list
EXTERNAL external name list
FORMAT ( [ format specification list ] )
IMPLICIT implicit specification
INTENT ( intent specification ) :: dummy argument name list
INTRINSIC intrinsic procedure name list
OPTIONAL [ :: ] optional object list
PARAMETER ( named constant definition list )
POINTER [ :: ] pointer name list
PUBLIC [ [ :: ] module entity name list ]
PRIVATE[ [ :: ] module entity name list ]
SAVE[ [ :: ] saved object list ]
TARGET [ :: ] target name list
USE module name [ , rename list ]
USE  module name , ONLY : [ access list ]
type specification [ [ , attribute specification ] ... :: &
      object declaration list
```

25.10.11 Type specification

```
INTEGER [ ( [ KIND= ] kind parameter ) ]
REAL[ ( [ KIND= ] kind parameter ) ]
COMPLEX[ ( [ KIND= ] kind parameter ) ]
CHARACTER[ ( [ KIND= ] kind parameter ) ]
CHARACTER[ ( [ KIND= ] kind parameter ) ] &
   [ LEN= ] length parameter )
LOGICAL[ ( [ KIND= ] kind parameter ) ]
TYPE ( type name )
```

25.10.12 Attribute Specification

```
ALLOCATABLE
DIMENSION ( array specification )
```

```
EXTERNAL
INTENT ( intent specification )
INTRINSIC
OPTIONAL
PARAMETER
POINTER
PRIVATE
PUBLIC
SAVE
TARGET
```

25.10.13 Executable construct

```
action statement
case construct
do construct
if construct
where construct
```

25.10.14 Action statement

```
ALLOCATE ( allocation list ) [ ,STAT= scalar integer variable ] )
CALL subroutinename [ ( [ actual argument specification list] ) ]
CLOSE ( close specification list )
CYCLE [ do construct name ]
DEALLOCATE( name list ) [ , STAT= scalar integer variable ] )
ENDFILE external file unit
EXIT [ do construct name ]
GOTO label
IF ( scalar logical expression ) action statement
INQUIRE ( inquire specification list ) [ output item list ]
NULLIFY ( pointer object list )
OPEN ( connect specification list )
PRINT format [ , output item list ]
READ (i/o control specification list ) [ input item list ]
READ format [ , output item list ]
RETURN [ scalar integer expression ]
REWIND ( position specification list )
STOP [ access code ]
WHERE ( array logical expression ) array assignment expression
WRITE ( i/o control specification list ) [ output item list ]
pointer variable => target expression
variable = expression
```

26
Case Studies

The good teacher is a guide who helps others to dispense with his services.
R. S. Peters, *Ethics and Education*.

Aims

The aims of this chapter are to look at several complete examples highlighting a variety of aspects of the use of Fortran 90:–

- the solution of a set of ordinary differential equations using the Runge-Kutta-Merson method, with the use of a procedure as a parameter, and the use of work arrays;
- generic procedures;
- a function that returns a variable length array;
- operator and assignment overloading;
- diagonal extraction of a matrix
- modules and packing;

26 Case Studies

This chapter looks at more realistic case studies of the use of Fortran 90 and its new features. There are examples of:–

- the solution of a set of ordinary differential equations using the Runge-Kutta-Merson method, with the use of a procedure as a parameter, and the use of work arrays;
- the construction of generic procedures in Fortran 90. Many of the internal functions will take arguments of a variety of data types and return a result of the same type, e.g. SINE will take an integer argument, real argument of whatever precision, complex argument and return the appropriate result.
- operator and assignment overloading and the use of a MODULE PROCEDURE;
- the extraction of the diagonal elements of a matrix;

The examples have been chosen to highlight the better features of Fortran 90 and what is possible with a modern language.

26.1 Example 1 – Solving a System of First Order Ordinary Differential Equations using Runge-Kutta-Merson

Simulation and mathematical modelling of a wide range of physical processes often leads to a system of ordinary differential equations to be solved. Ordinary differential equations also occur when approximate techniques are applied to more complex problems. We will restrict ourselves to a class of ordinary differential equations called initial value problems. These are systems for which all conditions are given at the same value of the independent variable. We will further restrict ourselves to first order initial value problems of the form:–

$$\begin{aligned} \dot{y}_1 &= f_1(\underline{y}, t) \\ \dot{y}_2 &= f_2(\underline{y}, t) \\ \cdot &= \cdot \\ \dot{y}_n &= f_n(\underline{y}, t) \end{aligned}$$

or

$$\underline{\dot{y}} = \underline{f}(\underline{y}, t) \qquad (1)$$

with initial conditions

$$\underline{y}(t_0) = \underline{y}_0$$

where

$$\underline{y} = \begin{pmatrix} y_1 \\ \cdot \\ y_n \end{pmatrix} \qquad \underline{f} = \begin{pmatrix} f_1 \\ \cdot \\ f_n \end{pmatrix} \qquad \underline{y}_0 = \begin{pmatrix} y_1(t_0) \\ \cdot \\ y_n(t_0) \end{pmatrix}$$

Case Studies

If we have a system of ordinary differential equations of higher order then they can be reformulated to a system of order one. See the NAG library documentation for solving ordinary differential equations.

One well known class of methods for solving initial value ordinary differential equations is Runge Kutta. In this example we have coded the Runge Kutta Merson algorithm which is a fourth order method and solves (1) from a point t = A to a point t = B.

It starts with a step length $h = (B - A)/100$ and includes a local error control strategy such that the solution at t+h is accepted if $|error\ estimate| <\ user\ defined\ tolerance$.

If this isn't satisfied the step length h is halved and the solution attempt is repeated until the above is satisfied or the step length is too small and the problem left unsolved. Assuming the error criterion was satisfied the algorithm progresses with a suitable step length solving the equations at intermediate points until the end point B is reached. For a full discussion of the algorithm and the error control mechanism used see *Numerical Methods in Practice*, Tim Hopkins and Chris Phillips.

```
MODULE Precisions
INTEGER,PARAMETER:: Long=SELECTED_REAL_KIND(15,307)
END MODULE Precisions
SUBROUTINE Runge_Kutta_Merson(Y,FUN,IFAIL,N,A,B,Tol)
!
! Runge-Kutta-Merson method for the solution of a system of N
! 1st order initial value ordinary differential equations.
! The routine tries to integrate from T=A to T=B with
! initial conditions in Y, subject to the condition that the
! Absolute Error Estimate <= Tol. The step length is
! adjusted automatically to meet this condition. If the
! routine is successful it returns with IFAIL = 0, T=B and
! the solution in Y.
!
USE Precisions
!
IMPLICIT NONE
! Define arguments
!
REAL (Long),INTENT(INOUT):: Y(:)
REAL(Long), INTENT(IN)::A,B,Tol
INTEGER,INTENT(IN)::N
INTEGER,INTENT(OUT)::IFAIL
!
INTERFACE
   SUBROUTINE FUN(T,Y,F,N)
```

```
         USE Precisions
         IMPLICIT NONE
         REAL(Long),INTENT(IN),DIMENSION(:)::Y
         REAL(Long),INTENT(OUT),DIMENSION(:)::F
         REAL(Long),INTENT(IN)::T
         INTEGER,INTENT(IN)::N
       END SUBROUTINE FUN
  END INTERFACE
  !
  ! Local variables
  !
  REAL(Long), DIMENSION(1:SIZE(Y)):: &
                S1,S2,S3,S4,S5,New_Y_1,New_Y_2,Error
  REAL(Long)::T,H,H2,H3,H6,H8,Factor=1.E-2_Long
  REAL(Long)::Smallest_step=1.E-6_Long,Max_Error
  INTEGER::No_of_steps=0
  !
       IFAIL=0
  !
  ! Check input parameters
  !
       IF(N <= 0 .OR. A == B .OR. Tol <= 0.0) THEN
          IFAIL = 1
          RETURN
       ENDIF
  !
  ! Initialize T to be start of interval and
  ! H to be 1/100 of interval
       T=A
       H=(B-A)/100.0_Long
       DO                                    ! Beginning of Repeat loop
          H2=H/2.0_Long
          H3=H/3.0_Long
          H6=H/6.0_Long
          H8=H/8.0_Long
  !
  ! Calculate S1,S2,S3,S4,S5
  !
          CALL FUN(T,Y,S1,N)                 ! S1=F(T,Y)

          New_Y_1=Y+H3*S1
          CALL FUN(T+H3,New_Y_1,S2,N)        ! S2 = F(T+H/3,Y+H/3*S1)
```

```
      New_Y_1=Y+H6*S1+H6*S2
      CALL  FUN(T+H3,New_Y_1,S3,N)         ! S3=F(T+H/3,Y+H/6*S1+H/6*S2)

      New_Y_1=Y+H8*(S2+3.0_Long*S3)
      CALL  FUN(T+H2,New_Y_1,S4,N)         ! S4=F(T+H/2,Y+H/8*(S2+3*S3))

      New_Y_1=Y+H2*(S1-3.0_Long*S3+4.0_Long*S4)
      CALL  FUN(T+H,New_Y_1,S5,N)          ! S5=F(T+H,Y+H/2*(S1-3*S3+4*S4))
!
! Calculate values at T+H
!
      New_Y_1=Y+H6*(S1+4.0_Long*S4+S5)
      New_Y_2=Y+H2*(S1-3.0_Long*S3+4.0*S4)
!
! Calculate error estimate
!
      Error=ABS(0.2_Long*(New_Y_1-New_Y_2))
      Max_Error=MAXVAL(Error)
      IF(Max_Error < Tol) THEN
!
! Halve step length and try again
!
         IF(ABS(H2) < Smallest_step) THEN
            IFAIL = 2
            RETURN
         ENDIF
         H=H2
      ELSE
!
! Accepted approximation so overwrite Y with Y_new_1,
! and T with T+H
!
         Y=New_Y_1
         T=T+H
!
! Can next step be doubled?
!
         IF(Max_Error*Factor < Tol)THEN
            H=H*2.0_Long
         ENDIF
!
```

```
! Does next step go beyond interval end B, if so set H = B-T
!
            IF(T+H > B) THEN
                H=B-T
            ENDIF
            No_of_steps=No_of_steps+1
        ENDIF
        IF(T >= B) EXIT                    ! End of repeat loop
    END DO
END SUBROUTINE Runge_Kutta_Merson
```

A main program to use this subroutine is of the form:

```
PROGRAM Odes
USE Precisions
IMPLICIT NONE
REAL(Long),Dimension(:),Allocatable::Y
REAL(Long)::A,B,Tol
INTEGER::N,IFAIL,All_stat
INTERFACE
    SUBROUTINE Runge_Kutta_Merson(Y,FUN,IFAIL,N,A,B,Tol)
        USE Precisions
        IMPLICIT NONE
        REAL(Long),INTENT(INOUT) :: Y(:)
        REAL(Long),INTENT(IN)::A,B,Tol
        INTEGER,INTENT(IN)::N
        INTEGER,INTENT(OUT)::IFAIL
        INTERFACE
            SUBROUTINE FUN(T,Y,F,N)
                USE Precisions
                IMPLICIT NONE
                REAL(Long), INTENT(IN),DIMENSION(:)::Y
                REAL(Long), INTENT(OUT),DIMENSION(:)::F
                REAL(Long), INTENT(IN)::T
                INTEGER,INTENT(IN)::N
            END SUBROUTINE FUN
        END INTERFACE
    END SUBROUTINE Runge_Kutta_Merson
END INTERFACE
!
```

Case Studies

```
   INTERFACE
      SUBROUTINE Fun1(T,Y,F,N)
      USE Precisions
      IMPLICIT NONE
      REAL(Long), INTENT(IN),DIMENSION(:)::Y
      REAL(Long), INTENT(OUT),DIMENSION(:)::F
      REAL(Long),INTENT(IN):: T
      INTEGER, INTENT(IN):: N
      END SUBROUTINE Fun1
   END INTERFACE
!
   PRINT *,'Input no of equations'
   READ*,N
!
! Allocate space for Y - checking to see that it
! allocates properly
!
   ALLOCATE(Y(1:N),STAT=All_stat)
   IF(All_stat /= 0) THEN
      PRINT*, 'Not enough memory, array Y is not allocated'
      STOP
   ENDIF
   PRINT *,' Input start and end of interval over'
   PRINT *,' which equations to be solved'
   READ *,A,B
   PRINT *,"Input ic's"
   READ *,Y(1:N)
   PRINT *,'Input Tolerance'
   READ *,Tol
   PRINT *,'At T= ',A
   PRINT *,'Initial conditions are :',Y(1:N)
   CALL Runge_Kutta_Merson(Y,Fun1,IFAIL,N,A,B,Tol)
   IF(IFAIL /= 0) THEN
      PRINT *,'Integration stopped with IFAIL = ',IFAIL
   ELSE
      PRINT *,'at T= ',B
      PRINT*,'Solution is: ',Y(1:N)
   ENDIF
END PROGRAM Odes
```

Consider trying to solve the following system of first order ordinary differential equations:–

$\dot{y}_1 = \tan y_3$

$$\dot{y}_2 = \frac{-0.032 \tan y_3}{y_2} - \frac{0.02\, y_2}{\cos y_3}$$

$$\dot{y}_3 = -\frac{0.032}{y_2^2}$$

over an interval $t = 0.0$ to $t = 8.0$ with initial conditions $y_1 = 0 \quad y_2 = 0.5 \quad y_3 = \frac{\pi}{5}$

The user supplied subroutine is:–

```
Subroutine Fun1(T,Y,F,N)
   USE Precisions
   IMPLICIT NONE
   REAL(Long),INTENT(IN),DIMENSION(:)::Y
   REAL(Long),INTENT(OUT),DIMENSION(:)::F
   REAL(Long),INTENT(IN)::T
   INTEGER,INTENT(IN)::N
!
   F(1)=TAN(Y(3))
   F(2)=-0.032_Long*F(1)/Y(2)-0.02_Long*Y(2)/COS(Y(3))
   F(3)=-0.032_Long/(Y(2)*Y(2))
END SUBROUTINE Fun1
```

26.1.1 Note: Alternative form of the Allocate statement

In the main program Odes we have defined Y to be a deferred shape array, allocating it space after the variable N is read in. In order to make sure that enough memory is available to allocate space to array Y the ALLOCATE statement is used as follows:–

ALLOCATE(Y(1:N),STAT=All_stat)

If the allocation is successful variable All_stat returns zero otherwise it is given a processor dependent positive value. We have included code to check for this and the program stops if All_stat is not zero.

26.1.2 Note: Automatic arrays

The subroutine Runge_Kutta_Merson needs a number of local rank-one arrays S1,S2,S3,S4 and S5 for workspace, their shape and size being the same as the dummy argument Y. Fortran 90 supplies automatic arrays for this purpose and can be declared as:–

```
REAL(Long), DIMENSION (1:SIZE(Y)) :: S1, S2, S3, S4, S5
```

The size of automatic arrays can depend on the size of actual arrays, in our example they are the same shape and size as the dummy array Y, or on some other dummy arguments. Automatic arrays

Case Studies

are created when the procedure is called and destroyed when control passes back to the calling program unit. They may have different shapes and sizes with different calls to the procedure, because of this automatic arrays cannot be saved or initialised.

A word of warning should be given at this point. If there isn't enough memory available when an automatic array needs to be created problems will occur. There is no way of testing to see if an automatic array has been created successfully, unlike allocatable arrays. The general feeling is that even though they are nice, automatic arrays should be used with care and shouldn't be used in production code!

26.1.3 Note: Dummy Procedure Arguments

In order to make the use of the subroutine Runge_Kutta_Merson as general as possible, one of its dummy arguments (FUN) is the name of a subroutine which the user supplies with the actual system of ordinary differential equations they wish to solve. This means that the main program which calls Runge_Kutta_Merson passes as an actual argument the name of the subroutine containing the definition of the equations they wish to solve. In this example the subroutine is called Fun1. In order to do this in the main program Odes we have an interface block for the Runge_Kutta_Merson subroutine and within this interface block we have another interface block for the dummy routine FUN. Also in the main program we have an interface block for the actual subroutine Fun1.

26.2 Example 2 – Generic Procedures

There has always been a degree of support in Fortran for the concept of a generic procedure. Simplistically a procedure is generic if it can handle arguments of more than data type. This concept is one that is probably taken for granted with the intrinsic procedures.

With Fortran 90 we can now define our own generic procedures. The example we will use is based on the earlier one of sorting. In the original example the program worked with real data. In the example below we have extended the program to handle both integer and real data.

What is not obvious from our use of the internal procedures is that there will be specific procedures that handle each data type, i.e. if a function can take integer, real and complex arguments then there will be one implementation of that function for each data type, i.e. three separate functions.

In the example below we add the ability to handle integer data. This means that where we had:—

- read data
- sort data
- print data

and one subroutine to implement the above we now have **two** subroutines to do each of the above, one to handle integers and one to handle reals.

```
PROGRAM Generic_Example
! This is an example of the ability in Fortran 90
! to construct generic procedures. The earlier example
! of sorting is now extended to handle both integer
! and real arrays.
IMPLICIT NONE
INTEGER :: How_Many
CHARACTER (LEN=20) :: File_Name
INTEGER,ALLOCATABLE , DIMENSION(:) :: I_Data
REAL ,ALLOCATABLE , DIMENSION(:) :: R_Data

INTERFACE Read_Data
   SUBROUTINE RR(File_Name,Raw_Data,How_Many)
      IMPLICIT NONE
      CHARACTER (LEN=*) , INTENT(IN) :: File_Name
      INTEGER , INTENT(IN) :: How_Many
      REAL , INTENT(OUT) , DIMENSION(How_Many) :: Raw_Data
   END SUBROUTINE RR
   SUBROUTINE RI(File_Name,Raw_Data,How_Many)
      IMPLICIT NONE
      CHARACTER (LEN=*) , INTENT(IN) :: File_Name
      INTEGER , INTENT(IN):: How_Many
      INTEGER , INTENT(OUT) , DIMENSION(How_Many) :: Raw_Data
   END SUBROUTINE RI
END INTERFACE

INTERFACE Sort_Data
   SUBROUTINE RS(Raw_Data,How_Many)
      IMPLICIT NONE
      INTEGER , INTENT(IN) :: How_Many
      REAL , INTENT(INOUT) , DIMENSION(How_Many) :: Raw_Data
   END SUBROUTINE RS
   SUBROUTINE IS(Raw_Data,How_Many)
      IMPLICIT NONE
      INTEGER , INTENT(IN) :: How_Many
      INTEGER , INTENT(INOUT) , DIMENSION(How_Many) :: Raw_Data
   END SUBROUTINE IS
END INTERFACE

INTERFACE Print_Data
   SUBROUTINE PR(Raw_Data,How_Many)
      IMPLICIT NONE
```

Case Studies

```
         INTEGER , INTENT(IN) :: How_Many
         REAL, INTENT(IN) , DIMENSION(How_Many) :: Raw_Data
      END SUBROUTINE PR
      SUBROUTINE PI(Raw_Data,How_Many)
         IMPLICIT NONE
         INTEGER , INTENT(IN) :: How_Many
         INTEGER , INTENT(IN) , DIMENSION(How_Many) :: Raw_Data
      END SUBROUTINE PI
END INTERFACE

   PRINT * , ' How many data items are there?'
   READ  * , How_Many
   PRINT * , ' What is the file name?'
   READ '(A)',File_Name
   ALLOCATE(I_Data(How_Many))
   CALL Read_Data(File_Name,I_Data,How_Many)
   CALL Sort_Data(I_Data,How_Many)
   CALL Print_Data(I_Data,How_Many)
   PRINT *, ' Phase 1 ends, data written to file name ISORTED.DAT'
   DEALLOCATE(I_Data)

   PRINT * , ' How many data items are there?'
   READ  * , How_Many
   PRINT * , ' What is the file name?'
   READ '(A)',File_Name
   ALLOCATE(R_Data(How_Many))
   CALL Read_Data(File_Name,R_Data,How_Many)
   CALL Sort_Data(R_Data,How_Many)
   CALL Print_Data(R_Data,How_Many)
   PRINT *, ' Program ends, data written to file name RSORTED.DAT'
END PROGRAM Generic_Example


SUBROUTINE RR(File_Name,Raw_Data,How_Many)
IMPLICIT NONE
CHARACTER (LEN=*) , INTENT(IN) :: File_Name
INTEGER , INTENT(IN) :: How_Many
REAL , INTENT(OUT) , DIMENSION(How_Many) :: Raw_Data
INTEGER :: I
   OPEN(FILE=File_Name,UNIT=1)
   DO I=1,How_Many
      READ (UNIT=1,FMT=*) Raw_Data(I)
```

```
      ENDDO
END SUBROUTINE RR



SUBROUTINE RI(File_Name,Raw_Data,How_Many)
IMPLICIT NONE
CHARACTER (LEN=*) , INTENT(IN):: File_Name
INTEGER , INTENT(IN) :: How_Many
INTEGER , INTENT(OUT) , DIMENSION(How_Many) :: Raw_Data
INTEGER :: I
   OPEN(FILE=File_Name,UNIT=1)
   DO I=1,How_Many
      READ (UNIT=1,FMT=*) Raw_Data(I)
   ENDDO
END SUBROUTINE RI



SUBROUTINE RS(Raw_Data,How_Many)
IMPLICIT NONE
INTEGER , INTENT(IN) :: How_Many
REAL , INTENT(INOUT) , DIMENSION(How_Many) :: Raw_Data
   CALL QuickSort(1,How_Many)

CONTAINS

RECURSIVE SUBROUTINE QuickSort(L,R)
IMPLICIT NONE
INTEGER , INTENT(IN) :: L,R
INTEGER :: I,J
REAL :: V,T
IF (R > L) THEN
   V=Raw_Data(R)
   I=L-1
   J=R
   DO
      DO
         I=I+1
         IF (Raw_Data(I) >= V) EXIT
      END DO
      DO
         J=J-1
         IF (Raw_Data(J) <= V) EXIT
```

```
            END DO
            T=Raw_Data(I)
            Raw_Data(I)=Raw_Data(J)
            Raw_Data(J)=T
            IF (J <= I) EXIT
         END DO
         Raw_Data(J)=Raw_Data(I)
         Raw_Data(I)=Raw_Data(R)
         Raw_Data(R)=T
         CALL QuickSort(L,I-1)
         CALL QuickSort(I+1,R)
    END IF
    END SUBROUTINE QuickSort

END SUBROUTINE RS


SUBROUTINE IS(Raw_Data,How_Many)
IMPLICIT NONE
INTEGER , INTENT(IN) :: How_Many
INTEGER , INTENT(INOUT) , DIMENSION(How_Many) :: Raw_Data
   CALL QuickSort(1,How_Many)

CONTAINS

RECURSIVE SUBROUTINE QuickSort(L,R)
IMPLICIT NONE
INTEGER , INTENT(IN) :: L,R
INTEGER :: I,J
INTEGER :: V,T
IF (R > L) THEN
   V=Raw_Data(R)
   I=L-1
   J=R
   DO
      DO
         I=I+1
         IF (Raw_Data(I)>=V) EXIT
      END DO
      DO
         J=J-1
         IF (Raw_Data(J) <= V) EXIT
```

```
         END DO
         T=Raw_Data(I)
         Raw_Data(I)=Raw_Data(J)
         Raw_Data(J)=T
         IF (J <= I) EXIT
      END DO
      Raw_Data(J)=Raw_Data(I)
      Raw_Data(I)=Raw_Data(R)
      Raw_Data(R)=T
      CALL QuickSort(L,I-1)
      CALL QuickSort(I+1,R)
   END IF
   END SUBROUTINE QuickSort

END SUBROUTINE IS


SUBROUTINE PR(Raw_Data,How_Many)
IMPLICIT NONE
INTEGER , INTENT(IN) :: How_Many
REAL , INTENT(IN) , DIMENSION(How_Many) :: Raw_Data
INTEGER :: I
   OPEN(FILE='RSORTED.DAT',UNIT=2)
   DO I=1,How_Many
      WRITE(UNIT=2,FMT=*) Raw_Data(I)
   END DO
   CLOSE(2)
END SUBROUTINE PR


SUBROUTINE PI(Raw_Data,How_Many)
IMPLICIT NONE
INTEGER , INTENT(IN) :: How_Many
INTEGER , INTENT(IN) , DIMENSION(How_Many) :: Raw_Data
INTEGER :: I
   OPEN(FILE='ISORTED.DAT',UNIT=2)
   DO I=1,How_Many
      WRITE(UNIT=2,FMT=*) Raw_Data(I)
   END DO
   CLOSE(2)
END SUBROUTINE PR
```

26.3 Example 3 – A Function that returns a variable length array

```
PROGRAM Running
! this program has a function that returns an array
! that contains the running average for the array
! passed

IMPLICIT NONE

INTERFACE
   FUNCTION Running_Average(A,How_Many)
      IMPLICIT NONE
      INTEGER , INTENT(IN) :: How_Many
      REAL , DIMENSION (How_Many) , INTENT(IN) :: A
      REAL , DIMENSION(How_Many) :: Running_Average
   END FUNCTION Running_Average
END INTERFACE

INTERFACE Read_Data

   SUBROUTINE RR(File_Name,Raw_Data,How_Many)
      IMPLICIT NONE
      CHARACTER (LEN=*) , INTENT(IN) :: File_Name
      INTEGER , INTENT(IN) :: How_Many
      REAL , INTENT(OUT) , DIMENSION(How_Many) :: Raw_Data
   END SUBROUTINE RR

   SUBROUTINE RI(File_Name,Raw_Data,How_Many)
      IMPLICIT NONE
      CHARACTER (LEN=*) , INTENT(IN) :: File_Name
      INTEGER , INTENT(IN) :: How_Many
      INTEGER , INTENT(OUT) , DIMENSION(How_Many) :: Raw_Data
   END SUBROUTINE RI

END INTERFACE

INTEGER :: How_Many
CHARACTER (LEN=20) :: File_Name
REAL , ALLOCATABLE , DIMENSION(:) :: Raw_Data
REAL , ALLOCATABLE , DIMENSION(:) :: RA
INTEGER :: I
   PRINT * , ' How many data items are there?'
   READ * , How_Many
```

```
      PRINT * , ' What is the file name?'
      READ '(A)',File_Name
      ALLOCATE(Raw_Data(How_Many))
      ALLOCATE(RA(How_Many))
      CALL Read_Data(File_Name,Raw_Data,How_Many)
      RA=Running_Average(Raw_Data,How_Many)
      DO I=1,How_Many
         PRINT *,Raw_Data(I),'    ' ,RA(I)
      END DO
END PROGRAM Running

FUNCTION Running_Average(R,How_Many)
INTEGER , INTENT(IN) :: How_Many
REAL , INTENT(IN) , DIMENSION(How_Many) :: R
REAL , DIMENSION(How_Many) :: Running_Average
INTEGER :: I
REAL :: Sum=0.0
   DO I=1,How_Many
      Sum = Sum + R(I)
      Running_Average(I)=Sum/I
   END DO
END FUNCTION Running_Average
```

26.4 Example 4 – Operator and Assignment Overloading

It is sometimes convenient to extend the meaning of an operator and the assignment symbol beyond that provided by the language. This can be done in Fortran 90 using module procedures. The example below is based on moving round a three dimensional space.

```
MODULE T_Position
IMPLICIT NONE
TYPE Position
   INTEGER :: X
   INTEGER :: Y
   INTEGER :: Z
END TYPE Position


INTERFACE OPERATOR (+)
   MODULE PROCEDURE New_Position
END INTERFACE
```

Case Studies

```
    CONTAINS

    FUNCTION New_Position(A,B)
    TYPE (Position) ,INTENT(IN) :: A,B
    TYPE (Position) :: New_Position
       New_Position % X = A % X + B % X
       New_Position % Y = A % Y + B % Y
       New_Position % Z = A % Z + B % Z
    END FUNCTION New_Position

    END MODULE T_Position

    PROGRAM Operator_Overload
    USE T_Position
    IMPLICIT NONE
    TYPE (Position) :: A,B,C
       A%X=10
       A%Y=10
       A%Z=10
       B%X=20
       B%Y=20
       B%Z=20
       C=A+B
       PRINT *,A
       PRINT *,B
       PRINT *,C
    END PROGRAM Operator_Overload
```

We have extended the meaning of the addition operator so that we can write simple expressions in Fortran based on the addition operator and have our new position calculated using a user supplied function that actually implements the calculation of the new position.

26.5 Example 5: A Subroutine to Extract the Diagonal Elements of a Matrix

A common task mathematically is to extract the diagonal elements of a matrix. For example if:–

$$A = \begin{pmatrix} 21 & 6 & 7 \\ 9 & 3 & 2 \\ 4 & 1 & 8 \end{pmatrix}$$

The diagonal elements are (21 , 3 , 8)

This can be thought of as extracting an array section, but the intrinsic function PACK is needed. In its simplest form PACK (Array,Vector) packs an array, Array, into a rank one array, Vector, according to Array's array element order.

Here's a subroutine to demonstrate this:

```
SUBROUTINE Matrix_Diagonal (A, Diag, N)
   IMPLICIT NONE
   REAL, INTENT(IN), DIMENSION(:,:) ::A
   REAL, INTENT(OUT), DIMENSION(:) :: Diag
   INTEGER, INTENT(IN) ::N
   REAL, DIMENSION (1:SIZE(A,1)*SIZE(A,1)) :: Temp
!
! Subroutine to extract the diagonal elements of
! an N * N matrix A
!
      IF(SIZE(A,1) == N .AND. SIZE(A,2) == N) THEN
         Temp = PACK(A,.TRUE.)
         Diag=Temp(1:N*N:N+1)
      ELSE
         PRINT*,'The matrix A is not (N,N), where N = ',N
         PRINT*,'Diagonal elements not extracted'
      END IF
END SUBROUTINE Matrix_Diagonal
```

26.6 Modules and Packaging

The power of modules is demonstrated very clearly when we look at how we package and make available data types and operations on those data types. The earlier example in this chapter on operator and assignment overloading was quite straightforward. Let us consider the provision of a windowing interface, with both text and graphics modes. The following is one possible (very simple) solution:–

MODULE Window, with procedures

    Open_Window
    Close_Window
    Draw_Title
    Redefine_Window
    Place_On_Top
    Place_Underneath
    On_Top
    Up_Window

MODULE Text_Window, with procedures

    Open_Text_Window
    Close_Text_Window
    Redefine_Text_Window

Case Studies

    Assign_Font
    Assign_Restore_Procedure
    Assign_EOW_Action
    Draw_Title
    Draw_Line
    Scroll_Up
    Scroll_Down
    Get_Position
    Set_Position
    Read_from_Window
    Write_To_Window

MODULE Graphic_Window, with procedures

    Open_GraphicsWindow
    Close_Graphics_Mode
    Select_Graphics_Mode
    Redefine_Graphics_Window
    Set_Pen
    Turn_To
    Turn
    Move_To
    Move
    Circle
    Area
    Copy_Area
    etc

MODULE Mouse_and_Cursor, with procedures

    Get_Mouse
    Set_Mouse
    Read_Mouse
    Move_Cursor
    Erase_Cursor
    etc.

Modules are first choice to implement this kind of problem over a suite of functions and subroutines because they provide a very clean way of packaging the whole solution.

As Fortran 90 is a new language there is little published material at this time that looks at the implementation of larger scale applications Two more mature languages are Ada and Modula 2, and it is worth looking at some of the published books on these languages. The bibliography provides one or two pointers.

26.7 Problems

1. Type in the subroutine to extract the diagonal elements of a matrix. Write a main program to call it, and try running it with matrices of your own choice.

26.8 Bibliography

Hopkins T., Phillips C., *Numerical Methods In Practice, Using the NAG Library.* Addison Wesley.

- Good adjunct to the NAG library documentation for the less numerate user.

NAG, *NAG Fortran 77 Library, MARK 16*, NAG Ltd.

- A library of over 1,000 numerical and statistical routines written by numerical analysts and fully tested and well documented. The documentation that accompanies the library is a must for serious users. There are versions of the Mark 15 Fortran 77 library which have been compiled with Fortran 90 compilers and these can then be used with Fortran 90 programs. Note that it is not possible to use the Fortran 77 library from a Fortran 90 program, for a variety of reasons.

NAG, NAG *fl90*, Release 1, NAG Ltd.

- This is a complete revamp of the Fortran 77 library using many of the new features of Fortran 90. It is currently a subset of the Fortran 77 library. There are 141 documented procedures grouped into 44 modules.

Tremblay J-P., deDourek J.M., Programming in Modula 2, McGraw Hill.

- One chapter looks at the provision of a *String* module as an addition to Modula 2, to extend the range of problems that can be solved in Modula 2.

Weiner R., Sinovec R., *Software Engineering with Modula 2 and Ada*, Wiley.

- Contains a number of examples that look at the use of modular programming, and its implementation in Ada and Modula 2. There is a complete example of a spelling checker implemented in both Ada and Modula 2.

27 Converting from Fortran 77

Twas brillig, and the slithy toves
did gyre and gimble in the wabe;
All mimsy were the borogoves,
And the mome raths outgrabe.

Lewis Carroll.

Aims

The aims of this chapter are to provide some simple guidelines for the conversion of existing Fortran 77 code to Fortran 90. In particular to emphasise the better features that exist in Fortran 90 in many areas.

- decremented features – candidates for deletion in future standards:–

    arithmetic if; real and double precision DO control variables; shared DO termination; alternate RETURN; PAUSE statement; ASSIGN and assigned GOTO statements; assigned FORMAT statements; H editing;

- better alternatives – potential candidates for removal from future standards

    double precision; fixed format; implicit typing; COMMON; EQUIVALENCE; block data; entry statement; statement functions; computed GOTO; alternate RETURN; INCLUDE; EXTERNAL, assumed size array arguments;

27 Converting from Fortran 77

The aim here is to provide the existing Fortran 77 programmer (and in particular the person with legacy code) with some simple guidelines for conversion.

The first thing that one must have is a thorough understanding of the newer, better language features of Fortran 90. It is essential that the material in the earlier chapters of this book be covered, and some of the problems attempted. This will provide a feel for Fortran 90.

The second thing one must have is a thorough understanding of the language constructs used in this legacy code. Use should be made of the compiler documentation for whatever Fortran 77 compiler you are using as this will provide the detailed (often system specific) information required. The recommendations below are therefore brief.

It is possible to move gradually from Fortran 77 to Fortran 90. In many cases existing code can be quite simply recompiled by a suitable choice of compiler options. This enables us to mix and match old and new in one program. This process is likely to highlight non-standard language features in your old code. There will inevitably be some problems here.

The first thing to consider is what the standard says. The standard identifies two kinds of decremented features. These are deleted and obsolescent features. It is extremely unwise to consider the long term use of these features as they are candidates for removal from future standards.

27.1 Deleted Features

The list of deleted features for Fortran 90 is empty, i.e. there are none.

27.2 Obsolescent Features

The obsolescent features are those for which better methods are available. They are given below with alternatives.

27.2.1 Arithmetic IF

Use the IF statement.

27.2.2 Real and Double precision DO Control Variables

Use integer.

27.2.3 Shared DO termination and non ENDDO termination

Use an END DO.

27.2.4 Alternate RETURN

Use a CASE statement on return. An error code has to be returned.

Converting from Fortran 77

27.2.5 PAUSE Statement

System specific. Normally easily replaced with a suitable READ statement.

27.2.6 ASSIGN and assigned GOTO statements

Fortunately rarely used.

27.2.7 Assigned FORMAT statements

Use character arrays, arrays and constants.

27.2.8 H Editing

Use character edit descriptor.

27.3 Better Alternatives

Here we are looking at the new features of the Fortran 90 standard, and how we can replace our current coding practices with the better facilities that now exist.

- DOUBLE PRECISION - use KIND, see chapter 8, and examples throughout the book.
- fixed format - use free format
- implicit typing - use IMPLICIT NONE
- BLOCK DATA - use modules
- COMMON statement - use modules
- EQUIVALENCE

    Invariably the use of this feature requires considerable system specific knowledge. There will be cases where there have been extremely good reasons why this feature has been used, normally efficiency related. However with the rapid changes taking place in the power and speed of hardware these reasons are diminishing.

- assumed size / explicit shape dummy array arguments

    If a dummy argument is assumed size or explicit shape (the only ones available in Fortran 77) then the ranks of the actual argument and the associated argument don't have to be the same. With Fortran 90 arrays are now objects instead of a linear sequence of elements, as was the case with Fortran 77, and now for array arguments the fundamental rule is that actual and dummy arguments have the same rank and same extents in each dimension, i.e. the same shape, and this is done using assumed shape dummy array arguments. An interface block is mandatory for assumed shape arrays.

- ENTRY statement - use module plus USE statement
- statement functions - use internal function, see chapter 14.

- computed GOTO - use CASE statement, see chapter 15.
- alternate RETURN - use error flags on calling routine
- INCLUDE - use modules plus USE statement
- EXTERNAL statement for dummy procedure arguments

    Use explicit interface blocks everywhere. See chapter 26, section 26.1 This also provides argument checking and other benefits.

28 Miscellanea

'The time has come,' the Walrus said,
　'To talk of many things:
of shoes–and ships–and sealing wax–
　of cabbages – and kings–
And why the sea is boiling hot–
　And whether pigs have wings.'
Lewis Carroll, *Through the Looking Glass, and What Alice Found There.*

Experience is a dear teacher, but fools will learn at no other.

Aims

The aim of this chapter is to provide a summary of what has been covered and help put some of the material into context. In particular:–

- program development and software engineering;
- programming style – programs should be easy to read;
- programming style – programs should behave well;
- data structures;
- algorithms
- recursion and when not to use it;
- structured programming and the GOTO statement;
- efficiency, the space time trade off;
- simple debugging guidelines;
- numerical software sources;

There is much to be learnt concerning the discipline of programming that does appear rather nonsensical at first.

28 Miscellanea

By now it should be apparent that coding is only one small part of programming. A thorough knowledge of a programming language is however a prerequisite for long term success.

Fortran has, because of its history, a number of drawbacks in terms of its syntax and semantics. There are some features that are quite clumsy in comparison to other, more modern languages. However there are disadvantages too in starting afresh each time, and this is highlighted by the Algol W, Pascal, Modula, Modula 2, (Modula 3), Oberon, Oberon 2 development. Moving from one language to its successor requires code changes of varying degrees of effort.

28.1 Program Development and Software Engineering

When one first starts programming the phrase software engineering will for many people appear meaningless. The problems that one solves when learning a programming language are by necessity small and amenable to fairly rapid solution and hence don't require too much thought and serious planning. This means that many people will initially have a rather simplistic attitude and approach to programming in the real world.

Consider the following classification, from Fairley (1985), *Software Engineering Concepts*:–

Classification	Number of programmers	Time Scale	Lines of code
Trivial	1	1–4 weeks	500
Small	1	1–6 months	1K–5K
Medium	2–5	1–2 years	5K–50K
Large	5–20	2–3 years	50K–100K
Very large	100–1000	4–5 years	1M
Massive	2000–5000	5–10 years	1M–10M

The development of reliable and correct programs requires increasing degrees of planning and organisation as one moves from trivial, to small, to medium and beyond. It is worthwhile looking again at the earlier coverage of systems analysis and design to see what stages we need to go through in the process of software engineering.

There are some practical considerations that can be applied regarding our use of a programming language, in this case Fortran 90.

28.1.1 Modules

Modules have a very important role to play in software engineering. Their addition to Fortran has made the language a much better vehicle for the construction of large scale, reliable and easily maintainable programs. They enable us to adhere to the two guidelines of logical coherence and independence.

How big should a program unit be? The rather trite statement here is *small is beautiful*. Research has consistently shown that the time taken to test and debug a program unit grows exponentially with program size. Thus we have the following:-

program size	testing and debugging time
1 unit	1 unit
2 units	4 units
3 units	8 units
4 units	32 units
etc	

Experience also shows that most program units are between 60 and 120 lines of code. The reasons for this are to do with page size, rather than anything wonderfully technical. It is what we can fit on one or two pages, i.e. what we can examine and understand easily. If we have to flip back and fore then we quickly loose track of what is happening.

28.1.2 Programming Style – Programs should be Easy to Read

Programs have to be read and understood by both computer systems (the compiler) and humans. Your time is a very valuable resource. It therefore pays to develop a good programming style. Clarity, simplicity and consistency are of the essence.

- meaningful names, not too long, not too short;
- comments should clarify the meaning of the code;
- indentation and formatting should be used to help structure the program and make it easier to read and understand; pretty print utilities may be available to help tidy up code.

The aim is not a Nobel prize for literature, rather a style that adds rather than detracts from understanding the meaning of the program.

28.1.3 Programming Style – Programs should Behave Well

Programs that behave badly will not be used, and most people reading this will know programs that they have stopped using precisely because they don't behave well. There are a number of things to consider here:-

- robustness, e.g. a compiler should not abort at the first error, but rather continue until some error limit is reached, preferably under user control;

- input validation should be carried out, and the first criterion is legal input data. A second criteria is to allow correction of invalid data without forcing the user to have to retype everything. The phrase garbage in garbage out immediately springs to mind!
- defensive programming – carry out tests before undertaking an operation. This will trap many common run-time errors, e.g. inappropriate argument to a function like the tangent of 90 degrees;
- graceful degradation, e.g. ignore the one erroneous record in a file and carry on processing the rest of the file;
- stick to the standard – as was stated earlier this pays off in the protection of the investment in people, in portability and a known reference point. If you have to use non-standard features then learn fully what that entails. The *apparently* identical language extension in two compilers can quite legitimately produce completely different results. Let the buyer beware!

28.2 Data Structures

Data structures have an enormous role to play in programming. An acquaintance with the wide range of data structuring techniques that is available is another skill that needs to be developed.

28.3 Algorithms

Algorithms have an important role to play in programming. The ability to choose the most appropriate algorithm for a particular problem is another skill that needs to be developed.

28.4 Recursion

Recursion is a very powerful tool in problem solving. However there are occasions when the overhead of recursion can be a problem. In this case it may be necessary to attempt to remove the recursion. There are a number of books that address this issue.

28.5 Structured Programming and the GOTO Statement

Consider the task of painting the floor boards in a room. Imagine getting a brush and a can of varnish and starting at the door. At some stage we end up in in the corner of the room and we have no way of leaving the room without walking over the recently painted boards. No problem we GOTO the door.

Whilst the above may strike one as silly the intention is to highlight the problems that can be created by inadequate preparation and planning.

As Knuth makes clear in *Structured Programming with goto Statements* the real issue is structured programming, and structured programming is not the process of writing programs and then eliminating their goto statements. The comment by W.W. Peterson in this paper (regarding the teaching of PL/I, where he taught students to use the goto in exceptional circumstances) highlights this point .. *A disturbingly large percentage of the students ran into situations that require gotos, and sure enough, it was often because* **while** *didn't work well to their plan, but almost invariably because their plan was poorly thought out.*

We have left the coverage of the GOTO statement until the last chapter to try and get people to think of solving problems without its use. The more astute reader may have already noticed that we have several examples throughout the notes that use an implied goto. These are the CYCLE and EXIT statements in loops. There is nothing wrong with using a goto providing its meaning is clear.

The syntax of the statement is:–

```
GOTO label
```

where label is a string of one to five decimal digits. Any statement in a Fortran program may have a label, e.g.

```
GOTO 100
```

Commonly used with a logical IF statement, e.g.

```
IF (Error) GOTO 1000
```

28.6 Efficiency, space time trade off

Efficiency is important, but the first priority must always be program correctness. A good understanding of the capabilities of the language, combined with a reasonable breadth of knowledge in data structuring and algorithms is a necessary precursor to concerns regarding efficiency.

There is a simple trade off available with many problems, space versus time. This means that the program will require a large amount of memory to solve a problem, and this will be reflected in a time reduction. Memory cannot be infinite on any system, so there will be instances when a longer execution time has to be endured through lack of physical memory.

It will often be necessary therefore to have only part of the data in memory. We will have to use disk to store the rest of the data we are processing. In situations like this it is recommended that you use unformatted files. As was stated earlier they offer no loss of precision, and none of the overhead involved in formatting.

28.7 Program Testing

Simplistically this means choosing a suite of data set inputs that test the following:–

- one or more commonly occurring cases;
- limiting cases;
- pathological cases;

Just because it works on a small range of test data does not mean that there are no bugs.

28.8 Simple Debugging Techniques

The first choice is to switch on all error testing that can be provided by the compiler, and flag all non-standard language features. This will trap the most common error which is array indexing going beyond the declared bounds, for any array data type.

Use of interface blocks will trap another very common error which is parameter mismatch between the calling routine and called routine.

The simplest debugging technique is the inclusion of print statements to clearly identify where the problem lies, and the state of the variables of interest. There is no substitute for sitting down and working through a listing of the program with sample runs to try and determine where problems lie.

28.9 Software Tools

There will be a range of tools available that aid in the program development process for the hardware and software platform that you use.

28.9.1 Cross Referencing

Good compilers will offer cross reference compilation options. It is probable that stand alone tools in this area will eventually become available over the Internet.

28.9.2 Pretty print

It is likely that tools to aid in consistent program style will become available in the near future over the Internet. NAG for example offers a tool to pretty print Fortran 90 programs.

28.9.3 NAGWare f90 Tools

The following tools are currently available:

- **Polisher**: Pretty prints Fortran 90 code.
- **Declaration Standardiser**: Standardises declarations of variables and parameters.
- **Precision Standardiser**: Modifies software to use parameterised precision.
- **Name Changer**: Changes names.

Additional tools will be added with later releases.

28.10 Numerical Software Sources

This software exists in two main forms.

Firstly as coded algorithms, which can be obtained in a variety of source forms, which you as end user include in your own program and compile with your particular compiler. Some of these are verified, others are made available with no warranty!

Secondly as commercially pre-compiled libraries for a variety of compilers, operating systems and hardware platforms. These are subject to rigorous testing by the suppliers and very well documented.

ACM – TOMS Transactions on Mathematical Software

They publish mathematical software as part of their collected algorithms of the ACM, available on magnetic tape and diskette. Further information is available on the World Wide Web, URL, http://www.acm.org

These sources are subject to validation and corrections and improvements are published.

IMSL

A commercial numerical and statistical library of approximately six hundred routines. Fortran 90 versions are becoming available.

NAG

A commercial numerical and statistical library. A Fortran 77 library of over one thousand routines has been available for 20 years. This library has been recompiled with a variety of Fortran 90 compilers. A completely redesigned Fortran 90 library has already been released and is subject to rapid development. There is a World Wide Web site for NAG too, URL, http://www.nag.co

Netlib

A collection of mathematical software, papers and databases. The software is free but you use it at your own risk, no support offered, and their motto is, *Anything free comes with no guarantee!*

Anonymous ftp to:

ftp.netlib.org

and log on as anonymous and use your complete email address as password. The site is the U.S.A. but there are local sites as well to ease the network traffic. For details of local sites send electronic mail containing the following message: `send site from netlib` to `netlib@ornl.gov`

Netlib is also accessible via the World Wide Web, and the URL is

http://www.netlib.org/index.html

28.11 Coda

Good luck!

28.12 Bibliography: All sources (bar one) taken from comp.software-eng.

28.12.1 Software Engineering

The best place to start is on USENET news. One of the FAQs is a comprehensive post on reading material for software engineers, and it is maintained by professional software engineers. Current top five are in order below, the sixth is recommended reading

Pfleeger S., *Software Engineering: The Production of Quality Software*, Macmillan, 1991.

Pressman R., *Software Engineering: A Practitioner's Approach*, McGraw Hill, 1987.

Sage A., Palmer J.D., *Software Systems Engineering*, ?.

Ghezzi, Jayazeri, Mandrioli, *Fundamentals of Software Engineering*, Prentice Hall.

Berzins V., Luqi, *Software Engineering with Abstractions*, Addison Wesley, 1991.

Brooks Frederick P. Jr, *The Mythical Man Month: Essays of Software Engineering*, Addison Wesley.

28.12.2 Programming Style

Anand N., *Clarify Function!*, ACM SigPLAN Notices, 23(6), 69–79.

Henry S., *A Technique for Hiding Proprietary Details While Providing Sufficient Information for Researchers; or, do you Recognise this Well Known Algorithm?*, Journal of Systems and Software, 8(1), 3–11.

Brooks R., *Studying Programmer Behaviour Experimentally: The Problems of Proper Methodology*, Communications of the ACM, 23(4), 207–213.

28.12.3 Software Testing

Beizer Boris, *Software Testing Techniques*, Van Nostrand Reinhold, 1990.

Hetzel William C., *The Complete Guide to Software Testing*, 2nd edition, QED Information Services Inc, 1988.

Software Research Inc., *Testing Techniques Newsletter*, Software Research Inc.

28.12.4 Fun

Martin Gardner, *Lewis Carroll: The Annotated Alice*, Penguin.

- Martin Gardner has been writing articles on mathematics as fun for twenty years. Lewis Carroll, nee Charles Lutwidge Dodgson was a lecturer in mathematics at Christ Church College, Oxford.

Glossary

actual argument
: A value (variable, expression or procedure) passed from a calling program unit to a subprogram unit.

adjustable array
: An explicit shape array that is a dummy argument to a sub program.

algorithm
: Derived from the name of the 9th century Persian mathematician Abu Ja'far Mohammed ibn Musa al-Kuwarizmi (father of Ja'far Mohammed, son of Moses, native of Kuwarizmi), corrupted through western culture as Al-Kuwarizmi. Now a sequence of computations.

allocatable array
: An array that has the ALLOCATABLE attribute.

argument
: Exists in two forms, actual argument which is in the calling routine, and is one of a variable, expression or procedure, and dummy argument which is in the called routine.

argument association
: The process of matching up an actual argument and dummy argument during program execution.

array
: An array is a data structure where each scalar element has the same type and kind. An array may be up to rank 7. It may be referenced by element (via subscripts), by section or as a whole.

array constructor
: A mechanism used to initialise or give values to a one dimensional array. The RESHAPE function can then be used to handle rank 2 and above arrays.

array element
: A scalar item of an array. An array element is picked out by a subscript.

array element ordering
: the elements of an array, regardless of rank, form a linear sequence. The sequence is such that the subscripts along the first dimension vary most rapidly.

array section
: A part of an array. The actual set depends on the subscripts.

ASCII
: American Standard Code for Information Interchange. See Appendix A.

association
: The means by which an entity can be referenced by different names in one scoping unit, or one or more names in multiple scoping units.

assumed length dummy argument
: A dummy argument that inherits the length attribute of the actual argument.

assumed shape array
: A dummy argument that inherits the shape of the associated argument.

assumed size array
: A dummy array whose size is inherited from the associated actual argument.

attribute
: A property of a data type, and specified in a type declaration statement.

automatic array
: This is an explicit shape array that is a local variable in a sub-program unit.

bound
: The bounds of an array are the upper and lower limits of the index in each dimension.

character constant
: A constant that is a string of one or more printable ASCII characters, enclosed in ' (apostrophe) or " (quotation mark).

character string
: A sequence of one or more characters. These are contiguous.

collating sequence
: The order in which a set of characters is sorted by default. The standard does not require that a processor provide the ASCII encoding, but does require intrinsic functions that convert between the processor encoding and the ASCII encoding.

compilation unit
: One or more source files that are compiled to form one object file.

component
: Part of a derived type definition.

concatenate
: join together two or more character items using the character concatenation operator //.

conformable
: Two arrays are said to be conformable if they have the same shape.

constant
: a constant is a data object whose value cannot be changed There are two kinds in Fortran, one is obtained using the PARAMETER statement, the other is a literal constant in an expression, e.g. with the expression 4*ATAN(1) both 4 and 1 are literal constants. It may be a scalar or an array.

contiguous
: Normally applied to items that are adjacent in memory, e.g. characters in a character variable.

data entity
: A data object that has a specific type.

data object
: a data object is a constant, variable or part of a constant or variable.

data type
: For each data type there is the following:- 0. a name 1. a set of values from a domain; 2. a set of valid operations upon these values; 3. a display method There are five predefined data types in Fortran and these are integer, real, complex, character and logical. For integers the values are drawn from the domain of integer numbers, the valid operations are addition, subtraction, multiplication, division and exponentiation, and they are displayed as a sequence of digits.

declaration
: A declaration is a non-executable statement that specifies attributes of a program element, e.g. specifying the dimension of an array, and the type of a variable.

default kind
: The kind type parameter which is used for one of Fortran's base types (integer, real, complex, character or logical) if one is not specified.

deferred shape array
: An allocatable array or an array pointer. The bounds are specified with a colon, e.g. (:).

defined
: For a data object having a valid value.

derived type
: A data type that is user defined and not one of the five intrinsic types.

dimension
: An array can be from one to seven dimensioned inclusive. Also called the rank.

dummy argument
: A variable name that appears in the bracketed or parenthesised list following the procedure name. (e.g. function or subroutine name). Dummy arguments take on the actual values of the corresponding arguments in the calling routine.

elemental
: An operation that applies independently to each element in an array.

entity
: Rather vague term covering a constant, variable, program unit, etc.

exceptional values
: Normally restricted to real numbers and typically one of non-normalised numbers, infinity, not-a-number (NaN) values etc.

explicit interface
: A mechanism to make information available between the calling routine and the called routine. This information includes the names of the procedures, the dummy arguments, the attributes of the arguments, the attributes of functions, and the order of the arguments.

explicit shape array
: A named array that has its bounds specified in each dimension.

expression
: An expression is a sequence of operands and operators that specifies a computation.

extent
: The number of elements of one dimension of an array. Also called the size.

external subprogram
: An external subprogram is one that is global to the whole program.

function
: One of the two procedure mechanisms available in Fortran along with the subroutine. they effectively provide a way of invoking a computation by using the function name, and return a result. There is the concept of type and kind for the result.

function reference
: A function is invoked by the use of its name in an expression.

function result
: The result value(s) from invoking a function.

generic
: Simplistically the ability of a procedure to accept arguments of more than type. This facility is taken for granted with the intrinsic procedures, and the user can now create their own generic procedures.

global
: An entity that is available throughout the executable program. A global entity has global scope. See also scope and local scope.

host association
: The mechanism by which a module procedure, internal procedure or derived type definition accesses entities of the host.

implicit interface
: A procedure interface whose properties are not known within the scope of the calling routine.

inquiry function
: A function whose result depends on the properties of the argument.

interface block
: A sequence of statements starting with an INTERFACE statement and ending with an END INTERFACE statement.

interface body
: The sequence of statements in an interface block between either a FUNCTION or SUBROUTINE statement and the corresponding END statement

internal procedure
: A procedure that is contained within an internal subprogram. the program unit containing the internal procedure is called the host. the internal procedure is local to the host and inherits the host environment through host association.

intrinsic procedure
: One of the standard supplied procedures.

kind
: For each of the five Fortran types (integer, real, complex, logical and character) there is the concept of kind. For example for integers it is common to find 8 bit, 16 bit and 32 bit implementations. Each of these has an associated kind type.

For real and complex types this enables us to chose both the range and precision of the numbers we work with.

For characters we can chose between character sets, which is of considerable use for working with different languages.

kind type parameter
An integer value used to identify the kind of one of the five base types, see above.

language extension
Most compiler implementations will provide language extensions. These are NOT part of the standard, and make porting code suites between different hardware and software platforms difficult and sometimes impossible.

linker
A program that is normally the final stage in the process of going from Fortran source to executable.

local entity
An entity that is only available within the context of a sub-program.

main program
A program unit that contains a PROGRAM statement.

module
A program unit that contains specifications and definitions that other program units can access and use.

module procedure
A function or subroutine defined within a module

name
An entity with a program e.g. constant, variable, function result, procedure, program unit, dummy argument.

name association
This provides access to the same entity (either data or a procedure) from different scoping units by the same or different name

nesting
The placing of one entity within another, e.g. one loop within another or one subprogram within another.

nonexecutable statement
A language statement that describes program attributes, but does not cause any action when the program is executed.

object file
File created after successful compilation. Used by the linker to generate an executable.

parameter
Term used to describe two completely different things. 1. a named constant - and hence the PARAMETER attribute 2. more generally equivalent to argument

pointer
A data object that has the POINTER attribute.

pointer association
The association of a part of memory to a pointer by means of a target.

precision
The number of significant digits in a real number.

procedure
A function or subroutine.

procedure interface
The statements that specify the name of a procedure the characteristics of that procedure the name of the dummy arguments the attributes of the dummy arguments the generic identifier (optionally) for the procedure

program
A program is an entity that can be compiled and executed on its own. There must be at least a declaration block and execution block.

program unit
A main program or a subprogram. The subprogram can be a function, subroutine or module.

rank
: The rank of an array is the number of dimensions.

recursion
: A property of a function or subroutine, and it means that the function or subroutine references itself directly or indirectly.

reference
: a data object reference is the appearance of a named entity in an executable statement requiring the value of the object.

relational expression
: An expression containing one or more of the relational operators and operands of numeric or character type.

scalar
: A single data object of any type. A scalar has a rank of zero.

scalar variable
: A variable of scalar type

scope and scoping unit
: The part of a program in which a name has a defined meaning. The name may be a named constant, a variable, a function, a procedure, or dummy argument. The part of the program is one of a program unit or subprogram, a derived type definition or a procedure interface body. Scoping units cannot overlap, but one scoping unit may be contained in another. In the latter case we have an example of host association.

shape
: The rank and extents of an array.

shape conformance
: Generally means that two or more arrays have the same rank and extent.

size
: The total number of elements in an array — the product of the extents.

source file
: A file known to the operating system that contains the Fortran statements.

statement
: An instruction in a programming language, and they are normally classified as executable and nonexecutable.

stride
: The increment in subscript triplet.

structure
: Either a scalar data object of derived type or a composite entity containing one or more subcomponents.

sub-program
: A user written or supplied function or subroutine.

subroutine
: A user subprogram that is invoked with the CALL statement. It can return one value, many values or no value at all to the calling program through the arguments.

subscript
: a scalar integer expression used to select an element of an array

substring
: a contiguous set of characters in a string.

target
: A named data object associated with a pointer.

transformational function
: An intrinsic function that is not elemental or inquiry.

truncation
: For real numbers the approximation obtained by chopping off the fractional part of the number and working with the integer part.
  For character variables removing one or more characters from a string.

type declaration
: One of the nonexecutable statements in Fortran, and one of INTEGER, REAL, COMPLEX, CHARACTER, LOGICAL or TYPE.

underflow
: A condition where the result of an arithmetic expression is smaller than the minimum value in the range for that data type.

user-defined type
: A data type that is defined by the user and not one of the intrinsic types.

variable
: A data object that has an associated memory location, whose value can be changed during program execution. A variable may be a scalar or an array.

Standards

There is a coverage of the standard that applies, where appropriate. For the more curious and inquisitive user they may be interested in the information held at **www.iso.ch** which is the International Standards Organisations world wide web page.

Ada: ISO/IEC 8652:1995

```
WITH TEXT_IO;USE TEXT_IO;
PROCEDURE Add IS
  X1 : FLOAT;
  X2 : FLOAT;
  Sum : FLOAT:=0.0;
BEGIN
PACKAGE FLT_IO IS NEW FLOAT_IO(FLOAT);
USE FLT_IO;
  PUT(" Type in the two numbers");
  GET(X1);
  GET(X2);
  Sum:=X1 + X2;
  NEW_LINE(1);
  PUT(X1);
  PUT(" + ");
  PUT(X2);
  PUT(" = ");
  PUT(Sum);
END Add;
```

Algol: Was ISO 1538, but this has been withdrawn.

Algol 68: No standard, but the major definition is in the Revised Report.

Apl: ISO 8485:1989

Basic: ISO/IEC 10279:1991, Full Standard; ISO 6373:1984, Minimal Conformance.

```
100 PRINT " Type in two numbers"
200 INPUT A, B
300 C = A + B
400 PRINT A, " + ", B, " = ", C
```

C: ISO/IEC 9899:1990

```
# include <stdio.h>
main()
{ float a,b,sum;
  printf(" Type in two numbers ");
  scanf("%f",&a);
```

```
scanf("%f",&b);
sum=a+b;
printf("%f",a);
printf(" + ");
printf("%f",b);
printf(" = ");
printf("%f",sum);
printf("\n"); }
```

C++: Draft at this time.

```
#include <iostream.hxx>
#include <math.h>
int main()
{   float a,b,sum;
  sum=0.0;
  cout < " Type in two numbers  " ;
  cin > a > b ;
  sum =  a + b ;
  cout < a < " + " < b < " = " < sum < "\n" ;
  return (0); }
```

Cobol: ISO/IEC 1989:1985

Fortran 90: ISO/IEC 1539:1990

```
PROGRAM Example
IMPLICIT NONE
REAL :: A
REAL :: B
REAL :: Sum=0.0
  PRINT * , ' Type in two numbers'
  READ * , A,B
  Sum = A + B
  PRINT * , A ,' + ', B , ' = ', Sum
END PROGRAM Example
```

ICON: No standard.

Lisp

Logo: No standard.

Modula 2: ISO/IEC Draft 10514

```
MODULE Example;
  FROM InOut IMPORT Write,WriteLn,WriteString;
  FROM InOut IMPORT ReadReal,WriteReal;
```

```
  VAR
    A,B : REAL;
    Sum : REAL;
BEGIN
  Sum := 0.0;
  WriteString(" Type in two numbers");
  WriteLn;
  ReadReal(A);
  ReadReal(B);
  Sum := A + B;
  WriteReal(A,10);
  WriteString(" + ");
  WriteReal(B,10);
  WriteString(" = ");
  WriteReal(Sum,10);
  WriteLn;
END Example.
```

Oberon: No standard.

Pascal: Pascal – ISO 7185:1990; Extended Pascal – ISO/IEC 10206: 1991

```
PROGRAM Example(INPUT,OUTPUT);
VAR
  A : REAL;
  B : REAL;
  Sum : REAL;
BEGIN
  WRITELN(' Type in two numbers');
  READLN(A,B);
  Sum := A + B;
  WRITELN(A, ' + ' , B ,' = ' , Sum)
END.
```

Postscript:

Prolog: ISO/IEC Draft 13211-1

SQL: ISO 9075:1992(E)

Simula: No international standard, but a Swedish one does exist.

Smalltalk:

Snobol:

ASCII Character Set

0	nul	32		64	@	96	`
1	soh	33	!	65	A	97	a
2	stx	34	"	66	B	98	b
3	etx	35	#	67	C	99	c
4	eot	36	$	68	D	100	d
5	enq	37	%	69	E	101	e
6	ack	38	&	70	F	102	f
7	bel	39	'	71	G	103	g
8	bs	40	(	72	H	104	h
9	ht	41	)	73	I	105	i
10	lf	42	*	74	J	106	j
11	vt	43	+	75	K	107	k
12	ff	44	,	76	L	108	l
13	cr	45	-	77	M	109	m
14	so	46	.	78	N	110	n
15	si	47	/	79	O	111	o
16	dle	48	0	80	P	112	p
17	dc1	49	1	81	Q	113	q
18	dc2	50	2	82	R	114	r
19	dc3	51	3	83	S	115	s
20	dc4	52	4	84	T	116	t
21	nak	53	5	85	U	117	u
22	syn	54	6	86	V	118	v
23	etb	55	7	87	W	119	w
24	can	56	8	88	X	120	x
25	em	57	9	89	Y	121	y
26	sub	58	:	90	Z	122	z
27	esc	59	;	91	[	123	{
28	fs	60	<	92	\	124	|
29	gs	61	=	93	]	125	}
30	rs	62	>	94	^	126	~
31	us	63	?1	95	_	127	del

Appendix C — Intrinsic Functions and Procedures

The following abbreviations and typographic conventions are used in this appendix:–

For argument type and result type

I	Integer
R	Real
C	Complex
N	Numeric (any of integer, real, complex)
L	Logical
P	Pointer
T	Target
DP	Double Precision
Char	Character, length = 1.
S	Character

For the class

E	elemental function
I	inquiry function
T	transformational function
S	subroutine

See chapter fourteen for more information on these classifications.

Arguments in italics, e.g.

ALL(Mask,*Dim*)

are optional arguments, i.e. Dim may be omitted in the example above.

Double Precision

Before Fortran 90 if you required real variables to have greater precision than the default real then the only option available was to declare them as double precision. With the introduction of kind types with Fortran 90 the use of double precision declarations is not recommended, and instead real entities with a kind type offering more than the default precision should be used.

Kind Optional Argument

There are several functions that have an optional argument Kind, e.g. AINT(A,*Kind*). If Kind is absent the result is the same kind type as the first argument, in this case A. If Kind is present the result has the kind type specified by this argument.

Result Type

When the result type is **as ...** then the result is not just the same type as the argument but also the same **kind**.

Function Name	Description	Argument Name	Type	Result	Class
• **ABS(A)**	Yields the absolute value unless A is complex, see below.	A	N	As argument	E

Note: If A is complex (x,y) then the functions returns $\sqrt{x^2+y^2}$

Example: R1=ABS(A)

• **ACHAR(I)**	Returns character in the ASCII character set.	I	I	Char	E

Example: C=ACHAR(I)

• **ACOS(X)**	Arccosine (inverse cosine)	X	R	As argument	E

Note: $|x| \leq 1$

Example: Y=ACOS(X)

• **ADJUSTL(String)**	Adjust string left, removing leading blanks and inserting trailing blanks.	String	S	As argument	E

Example: S=ADJUSTL(S)

• **ADJUSTR(String)**	Adjust string right, removing trailing blanks and inserting leading blanks.	String	S	As argument	E

Example: S=ADJUSTR(S)

• **AIMAG(Z)**	Imaginary part of complex argument	Z	C	As argument	E

Example: Y=AIMAG(Z)

• **AINT(A,***Kind***)**	Truncation	A Kind	R I	As A	E

Example: Y=AINT(Z) and when Z=0.3 Y=0, when Z=2.73 Y=2.0, when Z=-2.73 Y=-2.0

Function Name	Description	Argument Name	Type	Result	Class
ALL(Mask,*Dim***)**	Determines whether all values are true in Mask along dimension Dim.	Mask Dim	L I	L	T

Note: Dim must be a scalar in the range $1 \leq Dim \leq n$ where n is the rank of Mask. The result is scalar if Dim is absent or Mask has rank 1. Otherwise it works on the dimension Dim of Mask and the result is an array of rank n-1.

Example: T=ALL(M)

ALLOCATED(Array)	Returns true if array is allocated	Array	Any	L	I

Note: Array must be declared with the ALLOCATABLE attribute.

Example: IF (ALLOCATED(Array)) THEN ...

ANINT(A,*Kind***)**	Rounds reals, i.e. returns nearest whole number.	A Kind	R I	As A	E

Example: Z=ANINT(A), if A = 5.63 Z = 6, if A=-5.7 Z = -6.0

ANY(Mask,*Dim***)**	Determines whether any value is true in Mask along dimension Dim.	Mask Dim	L I	L	T

Note: Mask must be an array. The result is a scalar if Dim is absent or if Mask is of rank 1. Otherwise it works on the dimension Dim of Mask and the result is an array of rank n-1.

Example: T=ANY(A)

ASIN(X)	Arcsine	X	R	As argument	E

Example: Z=ASIN(X)

ASSOCIATED(Pointer, *Target***)**	Returns the association status of the pointer	Pointer Target	P P,T	L	I

Note:
. If Target is absent then the result is true if POINTER is associated with a target, otherwise false.
. If Target is present and is a target, the result is true if Pointer is currently associated with Target and false if is not.

Function Name	Description	Argument Name	Type	Result	Class

3. If Target is present and is a pointer, the result is true if both Pointer and Target are currently associated with the same target, and is false otherwise. If either Pointer or Target is disassociated the result is false.

Example: T=ASSOCIATED(P)

• **ATAN(X)**	Arctangent	X	R	As argument	E

Example: Z=ATAN(X)

• **ATAN2(Y,X)**	Arctangent of Y / X	Y X	R R	As arguments	E

Example: Z=ATAN2(Y,X)

• **BIT_SIZE(I)**	Returns the number of bits, as defined by the numeric model for integer numbers in chapter 8	I	I	As argument	I

Example: N_Bits=SIZE(I)

• **BTEST(I,Pos)**	Returns true if the bit is set in the integer argument at the position given by the second argument	I Pos	I I	L	E

Example: T=BTEST(I,Pos)

• **CEILING(A)**	Returns the smallest integer greater than or equal to the argument	A	R	I	E

Example: I=CEILING(A) If A=12.21 then I=13, if A=-3.16 then I=-3

• **CHAR(I,*Kind*)**	Returns the character in a given position in the processor collating sequence associated with the specified kind type parameter. Normally ASCII.	I Kind	I I	CHAR	E

Appendix C — Intrinsic Functions and Procedures — 317

Function Name	Description	Argument Name	Type	Result	Class

Example: C=CHAR(65) and for the ASCII character set C='A'.

| **CMPLX(X,Y,Kind)** | Converts to complex from integer, real and complex | X
Y
Kind | N
I, R
I | C | E |

Note:
1. If X is complex and Y is absent it is as if Y were present with the value AIMAG(X)
2. If X is not complex and Y is absent, it is as if Y were present with the value 0

Example: Z=CMPLX(X,Y)

| **CONJG(Z)** | Conjugate of a complex argument | Z | C | As Z | E |

Example: Z1=CONJG(Z)

| **COS(X)** | Cosine | X | R, C | As argument | E |

Note: The arguments of all trigonometric functions should be in radians, not degrees.

Example: A=COS(X)

| **COSH(X)** | Hyperbolic cosine | X | R | As argument | E |

Example: Z=COSH(X)

| **COUNT(Mask,Dim)** | Returns the number of true elements in Mask along dimension Dim. | Mask
Dim | L
I | I | T |

Note: Dim must be a scalar in the range $1 \leq Dim \leq n$ where n is the rank of Mask. The result is scalar if Dim absent or Mask has rank 1. Otherwise it works on the dimension Dim of Mask and the result is an array of rank n-1.

Example: N=COUNT(A)

| **CSHIFT(Array,Shift,Dim)** | Circular shift on a rank one array or rank one sections of higher rank arrays. | Array
Shift
Dim | Any
I
I | As Array | T |

Function Name	Description	Argument Name	Type	Result	Class

Note: Array must be an array, Shift must be a scalar if Array has rank one, otherwise it is an array of rank n-1, where n is the rank of Array. Dim must be a scalar with a value in the range $1 \leq Dim \leq n$.

Example: Array=CSHIFT(Array,10)

• **DATE_AND_TIME**(*Date, Time,Zone,Values*)	Returns the current date and time (compatible with ISO 8601:1988)	Date Time Zone Values	S S S I	N/A	S

Note:
1. Date is optional and must be scalar and of length 8 characters in order to return the complete value of the form CCYYMMDD where CC is the century, YY is the year, MM is the month and DD is the day. It is INTENT(OUT).
2. Time is optional and must be scalar and of length 10 characters in order to return the complete value of the form hhmmss.sss where hh is the hour, mm is the minutes and ss.sss is the seconds and milliseconds. It is INTENT(OUT).
3. Zone is optional and must be scalar and must be of length 5 characters in order to return the complete value of the form ±hhmm where hh and mm are the time differences with respect to Coordinated Universal Time in hours and minutes. It is INTENT(OUT).
4. Values is optional and a rank one array of size 8. It is INTENT(OUT). The values returned are as follows:–
Values(1) = the year
Values(2) = the month
Values(3) = the day
Values(4) = the time with respect to Coordinated Universal Time in minutes.
Values(5) = the hour (24 hour clock)
Values(6) = the minutes
Values(7) = the seconds
Values(8) = the milliseconds in the range 0–999

Example: CALL DATE_TIME(D,T,Z,V)

• **DBLE(A)**	Converts to double precision from integer, real, and complex	A	N	DP	E

Example: D=DBLE(A)

Function Name	Description	Argument Name	Type	Result	Class
DIGITS(X)	Returns the number of significant digits of the argument as defined in the numeric models for integer and reals in chapter 8	X	I,R	I	I

Example: I=DIGITS(X)

Function Name	Description	Argument Name	Type	Result	Class
DIM(X,Y)	Returns first argument minus minimum of the two arguments. X -MIN(X,Y)	X Y	I, R as X	As arguments	E

Example: Z=DIM(X,Y)

Function Name	Description	Argument Name	Type	Result	Class
DOT_PRODUCT(Vector_1, Vector_2)	Performs the mathematical dot product of two rank one arrays.	Vector_1 Vector_2	N, L As first argument	As arguments	T

Note:
1. If Vector_1 is of type integer or real the result has the value SUM(Vector_1*Vector_2)
2. If Vector_1 is complex the result has the value SUM(CONJG(Vector_1)*Vector_2)
3. If Vector_1 is logical the result has the value ANY(Vector_1 .AND. Vector_2)

Example: A=DOT_PRODUCT(X,Y)

Function Name	Description	Argument Name	Type	Result	Class
DPROD(X,Y)	Double precision product of two reals	X Y	R R	DP	E

Example: D=DPROD(X,Y)

Function Name	Description	Argument Name	Type	Result	Class
EOSHIFT(Array, Shift, Boundary,Dim)	End off shift of a rank 1 array or rank one section of a higher rank array	Array Shift Boundary Dim	Any I As Array I	As Array	T

Note: Array must be an array, Shift must be a scalar if Array has rank one, otherwise it is an array of rank n-1, where n is the rank of Array. Boundary must be scalar if Array has rank one, otherwise it must be either scalar or of rank n-1. Dim must be a scalar with a value in the range $1 \leq Dim \leq n$.

Example: A=EOSHIFT(A,Shift)

Function Name	Description	Argument Name	Type	Result	Class
• EPSILON(X)	Smallest difference between two reals of that kind. See chapter 8 and real numeric model	X	R	As argument	I

Example: Tiny=EPSILON(X)

Function Name	Description	Argument Name	Type	Result	Class
• EXP(X)	Exponential, e^x	X	R, C	As argument	E

Example: Y=EXP(X)

| • EXPONENT(X) | Returns the exponent component of the argument. See chapter 8 and the real numeric model | X | R | I | E |

Example: I=EXPONENT(X)

| • FLOOR(A) | Returns the greatest integer less than or equal to the argument | A | R | I | E |

Example: I=FLOOR(A) and when A=5.2 I has the value 5, when A=-9.7 I has the value -10

| • FRACTION(X) | Returns the fractional part of the real numeric model of the argument See chapter 8 and the real numeric model. | X | R | As X | E |

Example: F=FRACTION(X)

| • HUGE(X) | Returns the largest number for the kind type of the argument. See chapter 8 and the real and integer numeric models. | X | I,R | As argument | I |

Example: H=HUGE(X)

Function Name	Description	Argument Name	Type	Result	Class
IACHAR(C)	Returns the position of the character argument in the ASCII collating sequence	C	Char	I	E

Example: I=IACHAR('A') returns the value 65.

IAND(I,J)	Performs a logical AND on the arguments	I J	I As I	As arguments	E

Example: K=IAND(I,J)

IBCLR(I,Pos)	Clears one bit of the argument to zero.	I Pos	I I	As I	E

Note: $0 \leq Pos < BIT\_SIZE(I)$

Example: I=IBCLR(I,Pos)

IBITS(I,Pos,Len)	Returns a sequence of bits.	I Pos Len	I I I	As I	E

Note: $0 \leq Pos$ and $(Pos+Len) \leq BIT\_SIZE(I)$ and $Len \geq 0$

Example: Slice=IBITS(I,Pos,Len)

IBSET(I,Pos)	Sets one bit of the argument to one.	I Pos	I I	As I	E

Note: $0 \leq Pos < BIT\_SIZE(I)$

Example: I=IBSET(I,Pos)

ICHAR(C)	Returns the position of a character in the processor collating sequence associated with the kind type parameter of the argument. Normally the position in the ASCII collating sequence.	C	CHAR	I	E

Function Name	Description	Argument Name	Type	Result	Class

Example: I=ICHAR('A') would return the value 65 for the ASCII character set.

• **IEOR(I,J)**	Performs an exclusive OR on the arguments	I J	I As I	As I	E

Example: I=IEOR(I,J)

• **INDEX(String, Substring,***Back***)**	Locates one substring in another, i.e. returns position of Substring in character expression String.	String Substring Back	S S I	I	E

Note:
1. If Back is absent or present with the value .FALSE. then the function returns the start position of the first occurrence of the substring. If LEN(Substring) = 0 then one is returned.
2. If Back is present with the value .TRUE. then the function returns the start position of the last occurrence of the substring. If LEN(Substring) = 0 then the value (LEN(String) + 1) is returned.
3. If the substring is not found the result is zero.
4. If LEN(String) < LEN(Substring) the result is zero.

Example:
Where=INDEX(' Hello world Hello','Hello')
The result 2 is returned.
Where=INDEX(' Hello world Hello','Hello',.TRUE.)
The result 14 is returned.

• **INT(A,***Kind***)**	Converts to integer from integer, real, and complex	A Kind	N I	I	E

Example: I=INT(F)

• **IOR(I,J)**	Performs an inclusive OR on the arguments	I J	I As I	As I	E

Example: I=IOR(I,J)

• **ISHFT(I,Shift)**	Performs a logical shift. The bits of I are shifted by Shift positions.	I Shift	I I	As I	E

Note: $|Shift| \leq BIT\_SIZE(I)$

Appendix C   Intrinsic Functions and Procedures   323

Function Name	Description	Argument Name	Type	Result	Class

Example: I=ISHIFT(I,Shift)

| **ISHFTC(I,Shift,*Size*)** | Performs a circular shift of the rightmost bits. The Size rightmost bits of I are circularly shifted by Shift positions. | I
Shift
Size | I
I
I | I | E |

Note:
| Shift | ≤ Size
0 ≤ Size ≤ BIT_SIZE(I).
If Size is absent it is as if it were present with the value of BIT_SIZE(I).
If Shift is positive the shift is to the left.
If Shift is negative the shift is to the right.
If Shift is zero no shift is performed.

Example: I=ISHFTC(I,Shift,Size)

| **KIND(X)** | Returns the KIND type parameter of the argument. | X | Any | I | I |

Example: I=KIND(X)

| **LBOUND(Array,*Dim*)** | Returns the lower bounds for each dimension of the array argument or a specified lower bound. | Array
Dim | Any
I | I | I |

Note:
1 ≤ Dim ≤ n where n is the rank of Array. The result is scalar if Dim is present otherwise the result is an array of rank one and size n.
The result is scalar if Dim is present otherwise a rank one array and size n.

Example: I=LBOUND(Array)

| **LEN(String)** | Length of a character entity. | String | S | I | I |

Example: I=LEN(String)

Function Name	Description	Argument Name	Type	Result	Class
• LEN_TRIM(String)	Length of character argument less the number of trailing blanks	String	S	I	E
Example: I=LEN_TRIM(String)					
• LGE(String_1, String_2)	Lexically greater than or equal and this is based on the ASCII collating sequence.	String_1 String_2	S S	L	E
Example: L=LGE(S1,S2)					
• LGT(String_1,String_2)	Lexically greater than and this is based on the ASCII collating sequence.	String_1 String_2	S S	L	E
Example: L=LGT(S1,S2)					
• LLE(String_1,String_2)	Lexically less than or equal and this is based on the ASCII collating sequence.	String_1 String_2	S S	L	E
Example: L=LLE(S1,S2)					
• LLT(String_1, String_2)	Lexically less than and this is based on the ASCII collating sequence.	String_1 String_2	S S	L	E
Example: L=LLT(S1,S2)					
• LOG(X)	Natural logarithm, $\log_e x$	X	R, C	As argument	E
Example: Y=LOG(X)					
• LOG10(X)	Common logarithm, $\log_{10}$	X	R	As argument	E
Example: Y=LOG10(X)					

Function Name	Description	Argument Name	Type	Result	Class
LOGICAL(L, *Kind*)	Converts between different logical kind types, i.e. performs a type cast.	L Kind	L I	L	E

Example: L=LOGICAL(K,Kind)

Function Name	Description	Argument Name	Type	Result	Class
MATMUL(Matrix_1, Matrix_2)	Performs mathematical matrix multiplication of the array arguments	Matrix_1 Matrix_2	N,L As Matrix_1	As arguments	T

Note:
1. Matrix_1 and Matrix_2 must be arrays of rank one or two. If Matrix_1 is of numeric type so must Matrix_2.
2. If Matrix_1 has rank one, Matrix_2 must have rank two.
3. If Matrix_2 has rank one, Matrix_1 must have rank two.
4. The size of the first dimension of Matrix_2 must equal the size of the last dimension of Matrix_1.
5. If Matrix_1 has shape (n,m) and Matrix_2 has shape (m,k) the result has shape (n,k).
6. If Matrix_1 has shape (m) and Matrix_2 has shape (m,k) the result has shape (k).
7. If Matrix_1 has shape (n,m) and Matrix_2 has shape (m) the result has shape (n).

Example: R=MATMUL(M_1,M_2)

Function Name	Description	Argument Name	Type	Result	Class
MAX(A1,A2,*A3*,...)	Returns the largest value	A1 A2 A3 ...	I,R As A1 As A1 ...	As arguments	E

Example: A=MAX(A1,A2,A3,A4)

Function Name	Description	Argument Name	Type	Result	Class
MAXEPONENT(X)	Returns the maximum exponent. See chapter 8 and numeric models.	X	R	I	I

Example: I=MAXEXPONENT(X)

Function Name	Description	Argument Name	Type	Result	Class
MAXLOC(Array,*Mask*)	Determine the location of the first element of Array having the maximum value of the elements identified by Mask.	Array Mask	I,R L	I	T

Function Name	Description	Argument Name	Type	Result	Class

Note:
1. Array must be an array.
2. Mask must be conformable with Array
3. The result is an array of rank one and of size equal to the rank of Array.

Example: I=MAXLOC(Array)

In the above example if Array is a rank two array of shape (5,10) and the largest value is in position (2,1) the the result is the rank one array I with shape (2) and I(1)=2 and I(2)=1.

• **MAXVAL(Array,**Dim**,** **Mask)**	Returns the maximum value of the elements of Array along dimension Dim corresponding to the true elements of Mask.	Array Dim Mask	I,R I L	As argument	T

Note:

$1 \leq Dim \leq n$ where n is the rank of Array. The result is scalar if Dim is absent, or Array has rank one Otherwise the result is an array of rank n-1.

If Array has size zero then the result is the largest negative number supported by the processor for the corre sponding type and kind of Array.

Example:

MAXVAL((/1,2,3/)) returns the value 3.

MAXVAL(C,MASK=C < 0.0) returns the maximum of the negative elements of C.

For $B = \begin{pmatrix} 1 & 3 & 5 \\ 2 & 4 & 6 \end{pmatrix}$

MAXVAL(B,DIM=1) returns (2,4,6)

MAXVAL(B,DIM=2) returns (5,6)

• **MERGE(True,False, Mask)**	Chooses alternative values according to the value of a mask.	True False Mask	Any As True L	As True	E

Example: for $For\ True = \begin{pmatrix} 2 & 6 & 10 \\ 4 & 8 & 12 \end{pmatrix}$, $False = \begin{pmatrix} 1 & 5 & 9 \\ 3 & 7 & 11 \end{pmatrix}$ and $Mask = \begin{pmatrix} T & F & T \\ F & T & F \end{pmatrix}$ the result is $\begin{pmatrix} 2 & 5 & 10 \\ 3 & 8 & 11 \end{pmatrix}$

Function Name	Description	Argument Name	Type	Result	Class
MIN(A1,A2,*A3*,...)	Chooses the smallest value	A1 A2 A3,...	I, R As A1 As A1	As arguments	E

Example: Y=MIN(X1,X2,X3,X4,X5)

MINEXPONENT(X)	Returns the minimum exponent. See chapter 8 and numeric models	X	R	I	I

Example: I=MINEXPONENT(X)

MINLOC(Array,*Mask*)	Determine the location of the first element of Array having the minimum value of the elements identified by Mask.	Array Mask	I,R L	I	T

Note:
Array must be an array
Mask much be conformable with Array
The result is an array of rank one and of size equal to the rank of Array.

Example: I=MINLOC(Array)

In the above example if Array is a rank two array of shape (5,10) and the smallest value is in position (2,1) then the result is the rank one array I with shape (2) and I(1)=2 and I(2)=1.

MINVAL(Array, ***Dim,Mask*)**	Returns the minimum value of the elements of Array along dimension Dim corresponding to the true elements of Mask.	Array Dim Mask	I,R I L	As Array	T

Note: $1 \leq Dim \leq n$ where n is the rank of Array. The result is scalar if Dim is absent, or Array has rank one. Otherwise the result is an array of rank n-1.
If Array has size zero then the result is the largest negative number supported by the processor for the corresponding type and kind of Array.

Example:

MINVAL((/1,2,3/)) returns the value 1.

Function Name	Description	Argument Name	Type	Result	Class

MINVAL(C,MASK=C > 0.0) returns the minimum of the positive elements of C

For $B = \begin{pmatrix} 1 & 3 & 5 \\ 2 & 4 & 6 \end{pmatrix}$

MINVAL(B,DIM=1) returns (1,3,5)

MINVAL(B,DIM=2) returns (1,2)

• **MOD(A,B)**	Returns the remainder when first argument divided by second	A B	I, R As A	As arguments	E

Note: If B=0 the result is processor dependent. For B ≠ 0 the result is A - INT(A/B) * B

Example: R=MOD(A,B)
If A=8 and B=5 then R=3
If A=-8 and B=5 then R=-3
If A=8 and B=-5 then R=3
If A=-8 and B=-5 then R=-3

• **MODULO(A,B)**	Returns the modulo of the arguments	A B	I,R As A	As A	E

Note:
1. If B=0 then the result is processor dependent.
2. Integer A
The result is R where A= Q * B + R and Q is integer
for B>0, 0 ≤ R < B
for B < 0, B < R ≤ 0
3. Real A
The result is A - FLOOR(A/B) * B

Example: R=MODULO(A,B)
If A=8 and B=5 then R=3
If A=-8 and B=5 then R=2
If A=8 and B=-5 then R=-2
If A=-8 and B=-5 then R=-3

Function Name	Description	Argument Name	Type	Result	Class
MVBITS(From,F_Pos,Len, To,T_Pos)	Copies a sequence of bits from one data object to another	From F_Pos Len To T_Pos	I I I I I	N/A	S

Note:
From must be INTENT(IN).
F_Pos must be INTENT(IN), F_Pos ≥ 0, F_Pos+Len ≤ BIT_SIZE(From)
Len must be INTENT(IN), Len ≥ 0
To must be INTENT(INOUT)
T_Pos must be INTENT(IN), T_Pos ≥ 0, T_Pos + Len ≤ BIT_SIZE(To)
Example: CALL MVBITS(F,FP,L,T,TP)

Function Name	Description	Argument Name	Type	Result	Class
NEAREST(X,Next)	Returns the nearest different number. See chapter 8 and the real numeric model.	X Next	R R	As X	E

Example: N=NEAREST(X,Next)

Function Name	Description	Argument Name	Type	Result	Class
NINT(A,*Kind*)	Yields nearest integer	A Kind	R I	I	E

Note:
If A > 0, the result is INT(A+0.5)
If A ≤ 0, the result is INT(A-0.5)
Example: I=NINT(X)

Function Name	Description	Argument Name	Type	Result	Class
NOT(I)	Returns the logical complement of the argument	I	I	As I	E

Example: I=NOT(I)

Function Name	Description	Argument Name	Type	Result	Class
PACK(Array,Mask, *Vector*)	Packs an array into an array of rank one, under the control of a mask.	Array Mask Vector	Any L As Array	As Array	T

330   Intrinsic Functions and Procedures   Appendix

Function Name	Description	Argument Name	Type	Result	Class

Note:
1. Array must be an array.
2. Mask be conformable with Array.
3. Vector must have rank one and have at least as many elements as there are TRUE elements in Mask.
4. If Mask is scalar with the value TRUE Vector must have at least as many elements as there are in Array.
5. The result is an array of rank one.
6. If Vector is present the result size is that of Vector.
7. If Vector is not present the result size is t, the number of TRUE elements in Mask, unless Mask is scalar with a value TRUE in which case the result size is the size of Array.

Example: R=PACK(A,M)

• **PRECISION(X)**	Returns the decimal precision of the argument. See chapter 8 and numeric models	X	R, C	I	I

Example: I=PRECISION(X)

• **PRESENT(A)**	Returns whether an optional argument is present	A	Any	L	I

Note: A must be an optional argument of the procedure in which the PRESENT function reference appears.

Example: IF (PRESENT(X)) THEN ...

• **PRODUCT(Array,*Dim, Mask*)**	The product of all of the elements of Array along the dimension Dim corresponding to the TRUE elements of Mask.	Array Dim Mask	N I L	As Array	T

Note:
1. Array must be an array
2. $1 \leq Dim \leq n$ where n is the rank of Array
3. Mask must be conformable with Array
4. Result is scalar if Dim is absent, or Array has rank one, otherwise the result is an array of rank n-1

Example:
1. PRODUCT((/1,2,3/)) the result is 6.
2. PRODUCT(C,Mask=C > 0.0) forms the product of the positive elements of C.

Function Name	Description	Argument Name	Type	Result	Class

If $B = \begin{pmatrix} 1 & 3 & 5 \\ 2 & 4 & 6 \end{pmatrix}$

PRODUCT(B,DIM=1) is (2,12,30) and
PRODUCT(B,DIM=2) is (15,48)

Function Name	Description	Argument Name	Type	Result	Class
RADIX(X)	Returns the base of the numeric argument. See chapter 8 and numeric models.	X	I,R	I	I

Example: Base=RADIX(X)

RANDOM_NUMBER(X)	Returns one pseudo-random number or an array of pseudo-random numbers from the uniform distribution over the range $0 \leq x < 1$	X	R	N/A	S

Note: X is INTENT(OUT)

Example: CALL RANDOM_NUMBER(X)

RANDOM_SEED(Size, Put,Get)	Restarts (seeds) or queries the pseudo-random generator used by RANDOM_NUMBER	Size Put Get	I I I	N/A	S

Note:
Size is INTENT(OUT). It is set to the number N of integers that the processor uses to hold the value of the seed.
Put is INTENT(IN). It is an array of rank one and size $\geq$ N. It is used by the processor to set the seed value.
Get is INTENT(OUT). It is an array of rank one and size $\geq$ N. It is set by the processor to the current value of the seed.

Example: CALL RANDOM_SEED

Function Name	Description	Argument Name	Type	Result	Class
• RANGE(X)	Returns the decimal exponent range of the real argument. See chapter 8 and the numeric model representing the argument.	X	N	I	I

Example: I=RANGE(N)

Function Name	Description	Argument Name	Type	Result	Class
• REAL(A,*Kind*)	Converts to real from integer, real or complex	A Kind	N I	R	E

Example: X=REAL(A)

Function Name	Description	Argument Name	Type	Result	Class
• REPEAT(String, N_Copies)	Concatenates several copies of a string	String N_Copies	S I	S	T

Example: New_S=REPEAT(S,10)

Function Name	Description	Argument Name	Type	Result	Class
• RESHAPE(Source, Shape,*Pad,Order*)	Constructs an array of a specified shape from the elements of a given array	Source Shape Pad Order	Any I As Source I	As Source	T

Note:
1. Source must be an array. If Pad is absent or of size zero the size of Source must be ≥ PRODUCT(Shape)
2. Shape must be a rank one array and $0 \leq size < 8$
3. Pad must be an array.
4. Order must have the same shape as Shape and its value must be a permutation of (1,2,... ,n) where n is the size of Shape. If absent it is as if it were present with the value (1,2,...,n).
5. The result is an array of shape, Shape.

Example:
RESHAPE((/1,2,3,4,5,6/),(/2,3/)) has the value $\begin{pmatrix} 1 & 3 & 5 \\ 2 & 4 & 6 \end{pmatrix}$

RESHAPE((/1,2,3,4,5,6/) , (/2,4/) , (/0,0/) , (/2,1/)) has the value $\begin{pmatrix} 1 & 2 & 3 & 4 \\ 5 & 6 & 0 & 0 \end{pmatrix}$

Function Name	Description	Argument Name	Type	Result	Class
RRSPACING(X)	Returns the reciprocal of the relative spacing of model numbers near the argument value. See chapter 8 and real numeric model.	X	R	As X	E

Example: Z=RRSPACING(X)

SCALE(X,I)	Returns $X * b^I$ where b is the base in the model representation of X. See chapter 8 and the real numeric model.	X I	R I	As X	E

Example: Z=SCALE(X,I)

SCAN(String,Set,Back)	Scans a string for any one of the characters in a set of characters.	String Set Back	S As String L	I	E

Note:
1. The default is to scan from the left, and will only be from the right when Back is present and has the value TRUE.
2. Zero is returned if the scan fails.

Example: W=SCAN(String,Set)

SELECTED_INT_KIND(R)	Returns a value of the kind type parameter of an integer data type that represents all integer values n with $-10^R < n < 10^R$	R	I	I	T

Note:
1. R must be scalar
2. If a kind type parameter is not available then the value -1 is returned.

Example: I=SELECTED_INT_KIND(2)

Function Name	Description	Argument Name	Type	Result	Class
• SELECTED_REAL_KIND(*P,R*)	Returns a value of the kind type parameter of a real data type with decimal precision of at least P digits and a decimal exponent range of at least R.	P R	I I	I	T

Note:
1. P and R must be scalar.
2. The value -1 is returned if the precision is not available, the value -2 if the exponent range is not available and -3 if neither is available.

Example: I=SELECTED_REAL_KIND(P,R)

• SET_EXPONENT(X,I)	Returns the model number whose fractional part is the fractional part of the model representation of X and whose exponent part is I.	X I	R I	As X	E

Example: Exp_Part=SET_EXPONENT(X,I)

• SHAPE(Source)	Returns the shape of the array argument or scalar	Source	Any	I	I

Note:
1. Source may be array valued or scalar. It must not be a pointer that is disassociated or an allocatable array that is not allocated. It must not be an assumed size array.
2. The result is an array of rank one whose size is equal to the rank of Source.

Example: S=SHAPE(A(2:5,-1:1)) yields S=(4,3)

• SIGN(A,B)	Absolute value of A times the sign of B.	A B	I, R As A	As A	E

Example: A=SIGN(A,B)

• SIN(X)	Sine	X	R, C	As argument	E

Note: The argument is in radians.

Function Name	Description	Argument Name	Type	Result	Class

Example: Z=SIN(X)

| **SINH(X)** | Hyperbolic sine | X | R | As argument | E |

Example: Z=SINH(X)

| **SIZE(Array,*Dim*)** | Returns the extent of an array along a specified dimension or the total number of elements in an array. | Array
Dim | Any
I | I | I |

Note:
Array must be an array. It must not be a pointer that is disassociated or an allocatable array that is not allocated. If Array is an assumed size array Dim must be present with a value less than the rank of Array.
Dim must be scalar and in the range $1 \leq Dim \leq n$ where n is the rank of Array.
Result is equal to the extent of dimension Dim of Array, or if Dim is absent, the total number of elements of Array.

Example: A=SIZE(Array)

| **SPACING(X)** | Returns the absolute spacing of model numbers near the argument value. See chapter 8 and the real numeric model. | X | R | As X | E |

Example: S=SPACING(X)

| **SPREAD(Source,Dim,
N_Copies)** | Creates an array with an additional dimension, replicating the values in the original array | Source
Dim
N_Copies | Any
I
I | As Source | T |

Note:
Source may be array valued or scalar, with rank less than 7.
Dim must be scalar and in the range $1 \leq Dim \leq n+1$ where n is the rank of Source.
N_Copies must be scalar
The result is an array of rank n+1

Function Name	Description	Argument Name	Type	Result	Class

Example:

If A is the array (2,3,4) then SPREAD(A,DIM=1,NCOPIES=3) then the result is the array $\begin{pmatrix} 2 & 3 & 4 \\ 2 & 3 & 4 \\ 2 & 3 & 4 \end{pmatrix}$

• **SQRT(X)**	Square root	X	R, C	As argument	E

Example A=SQRT(B)

• **SUM(Array,***Dim,Mask***)**	Returns the sum of all elements of Array along the dimension Dim corresponding to the true elements of Mask	Array Dim Mask	N I L	As Array	T

Note:
1. Array must be an array
2. $1 \leq Dim \leq n$ where n is the rank of Array
3. Mask must be conformable with Array
4. Result is scalar if Dim is absent, or Array has rank one, otherwise the result is an array of rank n-1

Example:
1. SUM((/1,2,3/)) the result is 6.
2. SUM(C,Mask=C > 0.0) forms the arithmetic sum of the positive elements of C.
3. If $B = \begin{pmatrix} 1 & 3 & 5 \\ 2 & 4 & 6 \end{pmatrix}$
SUM(B,Dim=1) is (3,7,11)
SUM(B,Dim=2) is (9,12)

• **SYSTEM_CLOCK(***Count,Count_Rate,Count_Max***)**	Returns integer data from a real time clock	Count Count_Rate Count_Max	I I I	N/A	S

Note:
1. Count is INTENT(OUT) and is set to a processor dependent value based on the current value of the processor clock or to -HUGE(0) if there is no clock. $0 \leq Count \leq Count\_Max$
2. Count_Rate is INTENT(OUT) and it is set to the number of processor clock counts per second, or zero if there is no clock.

Function Name	Description	Argument Name	Type	Result	Class

. Count_max is INTENT(OUT) and is set to the maximum value that Count can have or to zero if there is no clock.

Example: CALL SYSTEM_CLOCK(C,R,M)

TAN(X)	Tangent	X	R	As argument	E

Note: X must be in radians.

Example: Y=TAN(X)

TANH(X)	Hyperbolic tangent	X	R	As argument	E

Example: Y=TANH(X)

TINY(X)	Returns the smallest positive number in the model representing numbers of the same type and kind type parameter as the argument	X	R	As X	I

Example: T=TINY(X)

TRANSFER(Source, Mold,*Size***)**	Returns a result with a physical representation identical to that of Source, but interpreted with the type and type parameters of Mold.	Source Mold Size	Any Any I	As Mold	T

Warning: A thorough understanding of the implementation specific internal representation of the data types involved is necessary for successful use of this function. Consult the documentation that accompanies the compiler that you work with before using this function.

TRANSPOSE(Matrix)	Transposes an array of rank two.	Matrix	Any	As argument	T

Note: Matrix must be of rank two. If its shape is (n,m) then the resultant matrix has shape (m,n)

Function Name	Description	Argument Name	Type	Result	Class

Example: For $A = \begin{pmatrix} 1 & 2 & 3 \\ 4 & 5 & 6 \\ 7 & 8 & 9 \end{pmatrix}$ TRANSPOSE(A) yields $\begin{pmatrix} 1 & 4 & 7 \\ 2 & 5 & 8 \\ 3 & 6 & 9 \end{pmatrix}$

- **TRIM(String)** — Returns the argument with trailing blanks removed — String — S — As String — T

Note: String must be a scalar

Example: T_S=TRIM(S)

- **UBOUND(Array,*Dim*)** — Returns all the upper bounds of an array or a specified upper bound. — Array / Dim — Any / I — I — I

Note:

$1 \leq Dim \leq n$ where n is the rank of Array. The result is scalar if Dim is present otherwise the result is an array of rank one and size n.

Result is a scalar if Dim is present otherwise is an array of rank one, and size n.

Example: Z=UBOUND(A)

- **UNPACK(Vector,Mask, Field)** — Unpacks an array of rank one into an array under the control of a mask. — Vector / Mask / Field — Any / L / As Vector — As Vector — T

Note:
1. Vector must have rank one. Its size must be at least t, where t is the number of true elements in Mask.
2. Mask must be array valued.
3. Field must be conformable with Mask. Result is an array with the same shape as Mask.

Example:

With $Vector = (1,2,3)$ and $Mask = \begin{pmatrix} F & T & F \\ T & F & F \\ F & F & T \end{pmatrix}$ and $Field = \begin{pmatrix} 1 & 0 & 0 \\ 0 & 1 & 0 \\ 0 & 0 & 1 \end{pmatrix}$ the result is $\begin{pmatrix} 1 & 2 & 0 \\ 1 & 1 & 0 \\ 0 & 0 & 3 \end{pmatrix}$

Function Name	Description	Argument Name	Type	Result	Class
VERIFY(String,Set,*Back*)	Verify that a set of characters contains all the characters in a string by identifying the position of the first character in a string of characters that does not appear in a given set of characters.	String Set Back	S As String L	I	E

Note:

1. The default is to scan from the left, and will only be from the right when Back is present and has the value TRUE.
2. The value of the result is zero if each character in String is in Set, or if String has zero length.

Example: I=VERIFY(String,Set)

YET IF HE SHOULD GIVE UP WHAT HE HAS BEGUN, AND AGREE TO MAKE US OR OUR KINGDOM SUBJECT TO THE KING OF ENGLAND OR THE ENGLISH, WE SHOULD EXERT OURSELVES AT ONCE TO DRIVE HIM OUT AS OUR ENEMY AND A SUBVERTER OF HIS OWN RIGHTS AND OURS, AND MAKE SOME OTHER MAN WHO WAS ABLE TO DEFEND US OUR KING; FOR, AS LONG AS BUT A HUNDRED OF US REMAIN ALIVE, NEVER WILL WE ON ANY CONDITIONS BE BROUGHT UNDER ENGLISH RULE. IT IS IN TRUTH NOT FOR GLORY, NOR RICHES, NOR HONOURS THAT WE ARE FIGHTING, BUT FOR FREEDOM - FOR THAT ALONE, WHICH NO HONEST MAN GIVES UP BUT WITH LIFE ITSELF.

QUEM SI AB INCEPTIS DIESISTERET, REGI ANGLORUM AUT ANGLICIS NOS AUT REGNUM NOSTRUM VOLENS SUBICERE, TANQUAM INIMICUM NOSTRUM ET SUI NOSTRIQUE JURIS SUBUERSOREM STATIM EXPELLERE NITEREMUR ET ALIUM REGEM NOSTRUM QUI AD DEFENSIONEM NOSTRAM SUFFICERET FACEREMUS. QUIA QUANDIU CENTUM EX NOBIS VIUI REMANSERINT, NUCQUAM ANGLORUM DOMINIO ALIQUATENUS VOLUMUS SUBIUGARI. NON ENIM PROPTER GLORIAM, DIUICIAS AUT HONORES PUGNAMUS SET PROPTER LIBERATEM SOLUMMODO QUAM NEMO BONUS NISI SIMUL CUM VITA AMITTIT.

from *'The Declaration of Arbroath'* c.1320. The English translation is by Sir James Fergusson.

Appendix E   Coded Text Extract   341

OH YABY NSFOUN, YAN DUBZY LZ DBUYLTUBFAJ BYYBOHNX GPDA FNUZNDYOLH YABY YAN SBF LZ B GOHTMN FULWOHDN DLWNUNX YAN GFBDN LZ BH NHYOUN DOYJ, BHX YAN SBF LZ YAN NSFOUN OYGNMZ BH NHYOUN FULWOHDN. OH YAN DLPUGN LZ YOSN, YANGN NKYNHGOWN SBFG VNUN ZLPHX GLSNALV VBHYOHT, BHX GL YAN DLMMNTN LZ DBUYLTUBFANUG NWLMWNX B SBF LZ YAN NSFOUN YABY VBG YAN GBSN GDBMN BG YAN NSFOUN BHX YABY DLOHDOXNX VOYA OY FLOHY ZLU FLOHY. MNGG BYYNHYOWN YL YAN GYPXJ LZ DBUYLTUBFAJ, GPDDNNXOHT TNHNUBYOLHG DBSN YL RPXTN B SBF LZ GPDA SBTHOYPXN DPSENUGLSN, BHX, HLY VOYALPY OUUNWNUNHDN, YANJ BEBHXLHNX OY YL YAN UOTLPUG LZ GPH BHX UBOH. OH YAN VNGYNUH XNGNUYG, YBYYNUNX ZUBTSNHYG LZ YAN SBF BUN GYOMM YL EN ZLPHX, GANMYNUOHT BH LDDBGOLHBM ENBGY LU ENTTBU; OH YAN VALMN HBYOLH, HL LYANU UNMOD OG MNZY LZ YAN XOGDOFMOHN LZ TNLTUBFAJ.

NAG have produced a version of their Mark 15 Fortran 77 library compiled with their NAGWare FTN90 compiler and the contents of this library are given below.

A02	Complex arithmetic
C02	Zeros of polynomials
C05	Roots of one or more transcendental equations
C06	Summation of series
D01	Quadrature
D02	Ordinary differential equations
D02M-D02N	Integrators for STIFF ordinary differential equations
D03	Partial differential equations
D04	Numerical differentiation
D05	Integral equations
E01	Interpolation
E02	Curve and surface fitting
E04	Minimising or maximising a function
F	Linear algebra
F01	Matrix operations including inversion
F02	Eigenvalues and eigenvectors
F03	Determinants
F04	Simultaneous linear equations
F05	Orthogonalisation
F06	Linear Algebra Support Routines
F07	Linear equations (LAPACK)
G01	Simple calculations on statistical data
G02	Correlation and regression analysis
G03	Multivariate methods
G04	Analysis of variance
G05	Random number generators
G07	Univariate Estimation
G08	Nonparametric statistics
G11	Contingency Table Analysis
G12	Survival analysis
G13	Time series analysis
H	Operations research

Appendix F   NAG   343

M01	Sorting
P01	Error trapping
S	Approximations of special functions
X01	Mathematical constants
X02	Machine constants
X03	Innerproducts
X04	Input/Output utilities
X05	Date and Time utilities

NAG have also produced a genuine Fortran 90 library, *fl90* Release 1, which takes full advantage of the facilities offered by Fortran 90. The library is organised into chapters. Within each chapter there are one or more modules and a module will contain one or more procedures. The organisation of the library is given below:–

Chapter 1	Utilities
Chapter 3	Special Functions
Chapter 5	Linear equations
Chapter 6	Eigenvalues and least-squares problems
Chapter 7	Transforms
Chapter 8	Curve and surface fitting
Chapter 9	Optimisation
Chapter 10	Nonlinear equations
Chapter 11	Quadrature
Chapter 12	Ordinary differential equations
Chapter 20	Statistical distribution functions
Chapter 21	Random number generation
Chapter 22	Basic descriptive statistics
Chapter 25	Correlation and regression analysis

Notes:

1. Missing chapters will be added at later releases of the library.

Syntax rules

This annex contains two parts. The first part is an extraction of all syntax rules and constraints in the order in which they appear in the International Standard. The second part is a cross reference with an entry for each terminal symbol and each nonterminal symbol in the syntax rules. The symbols are sorted alphabetically within three categories: nonterminal symbols that are defined, nonterminal symbols that are not defined, and terminal symbols.

Except for those ending with -name, the only undefined nonterminal symbols are letter, digit, special-character, and rep-char. As described in 1.5.2, the following rules are assumed. The letters "xyz" stand for any legal syntactic class phrase:

> **xyz-list** is xyz [, xyz] ...
> **xyz-name** is name
> **scalar-xyz** is xyz

Constraint: scalar-xyz must be scalar.

D.1 Syntax rules and constraints

Each of the following sections contains the syntax rules and constraints from one section of the standard; that is, Section D.1.1 contains the rules and constraints from Section 1 of the standard, D.1.2 contains those from Section 2, and so on. Note that Sections 1, 13, and 14 contain no syntax rules.

D.1.1 Overview

D.1.2 Fortran terms and concepts

R201 executable-program is program-unit
> [program-unit] ...

R202 program-unit is main-program
> or external-subprogram
> or module
> or block-data

R1101 main-program is [program-stmt]
> [specification-part]
> [execution-part]
> [internal-subprogram-part]
> end-program-stmt

R203 external-subprogram is function-subprogram
> or subroutine-subprogram

R1215 function-subprogram is function-stmt
> [specification-part]
> I execution-part]
> [internal-subprogram-part]
> end-function-stmt

R1219 subroutine-subprogram is subroutine-stmt
> [specification-part]
> [execution-part]
> [internal-subprogram-part]
> end-subroutine-stmt

R1104 module is module-stmt
> [specification-part]
> [module-subprogram-part]
> end-module-stmt

R1110 block-data is block-data-stmt
> [specification-part]
> end-block-data-stmt

R204 specification-part is [use-stmt] ...
> [implicit-part]
> [declaration-construct]

R205 implicit-part is [implicit-part-stmt] ...
> implicit-stmt

R206 implicit-part-stmt is implicit-stmt
> or parameter-stmt
> or format-stmt
> or entry-stmt

R207 declaration-construct is derived-type-def
> or interface-block
> or type-declaration-stmt
> or specification-stmt
> or parameter-stmt
> or format-stmt
> or entry-stmt
> or stmt-function-stmt

R208 execution-part is executable-construct
    [execution-part-construct] ...
R209 execution-part-construct is executable-construct
    or format-stmt
    or data-stmt
    or entry-stmt
R210 internal-subprogram-part is contains-stmt
    internal-subprogram
    [internal-subprogram]
R211 internal-subprogram is function-subprogram
    or subroutine-subprogram
R212 module-subprogram-part is contains-stmt
    module-subprogram
    [module-subprogram]
R213 module-subprogram is function-subprogram
    or subroutine-subprogram
R214 specification-stmt is access-stmt
    or allocatable-stmt
    or common-stmt
    or data-stmt
    or dimension-stmt
    or equivalence-stmt
    or external-stmt
    or intent-stmt
    or intrinsic-stmt
    or namelist-stmt
    or optional-stmt
    or pointer-stmt
    or save-stmt
    or target-stmt
R215 executable-construct is action-stmt
    or case-construct
    or do-construct
    or if-construct
    or where-construct
R216 action-stmt is allocate-stmt
    or assignment-stmt
    or backspace-stmt
    or call-stmt
    or close-stmt
    or computed-goto-stmt
    or continue-stmt
    or cycle-stm
    or deallocate-stmt
    or endfile-stmt
    or end-function-stmt
    or end-program-stmt
    or end-subroutine-stmt
    or exit-stmt
    or goto-stmt
    or if-stmt
    or inquire-stmt
    or nullify-stmt
    or open-stmt
    or pointer-assignment-stmt
    or print-stmt
    or read-stmt
    or return-stmt
    or rewind-stmt
    or stop-stmt
    or where-stmt
    or write-stmt
    or arithmetic-if-stmt
    or assign-stmt
    or assigned-goto-stmt
    or pause-stmt

Constraint: An execution-part must not contain an end-function-stmt, end-program-stmt, or end-subroutine-stmt.

D.1.3 Characters, lexical tokens, and source form

R301 character is alphanumeric-character
    or special-character
R302 alphanumeric-character is letter
    or digit
    or underscore
R303 underscore is _
R304 name is letter [alphanumeric-character]

Constraint: The maximum length of a name is 31 characters.

R305 constant is literal-constant
    or named-constant

R306 literal-constant is int-literal-constant
    or real-literal-constant
    or complex-literal-constant
    or logical-literal-constant
    or char-literal-constant
    or boz-literal-constant

R307 named-constant is name

R308 int-constant is constant

Constraint: int-constant must be of type integer.

R309 char-constant is constant

Constraint: char-constant must be of type character.

R310 intrinsic-operator is power-op
    or mult-op
    or add-op
    or concat-op
    or rel-op
    or not-op
    or and-op
    or or-op
    or equiv-op

R708 power-op is **

R700 mult-op is *
    or /

R710 add-op is +
    or -

R712 concat-op is //

R714 rel-op is .EQ.
    or .NE.
    or .LT.
    or .LE.
    or .GT.
    or .GE.
    or ==
    or /=

R719 not-op is .NOT.

R720 and-op is .AND.

R721 or-op is .OR.

R722 equiv-op is .EQV.
    or .NEQV.

R311 defined-operator is defined-unary-op
    or defined-binary-op
    or extended-intrinsic-op

R704 defined-unary-op is . letter [letter] ... .

R724 defined-binary-op is . letter [letter] ... .

R312 extended-intrinsic-op is intrinsic-operator

Constraint: A defined-unary-op and a defined-binary-op must not contain more than 31 letters and must not be the same as any intrinsic-operator or logical-literal-constant.

R313 label is digit [digit [digit [digit [digit]]]]

Constraint: At least one digit in a label must be non-zero.

D.1.4 Intrinsic and derived data types

R401 signed-digit-string is [sign] digit-string

R402 digit-string is digit [digit]

R403 signed-lit-literalconstant is [sign] int-literal-constant

R404 int-literal-constant is digit-string [_kind-param]

R405 kind-param is digit-string
    or scalar-int-constant-name

R406 sign is +
    or -

Constraint: The value of kind-param must be nonnegative.

Constraint: The value of kind-param must specify a representation method that exists on the processor.

R407 boz-literal-constant is binary-constant
    or octal-constant
    or hex-constant

Constraint: A boz-literal-constant may appear only in a DATA statement.

R408 binary-constant is B ' digit [digit] ...'
    or B " digit [digit] ..."

Constraint: digit must have one of the values 0 or 1.

R400 octal-constant is O ' digit [digit] ...'
O " digit [digit] ..."

Constraint: digit must have one of the values 0 through 7.

R410 hex-constant is Z ' hex-digit [hex-digit] ...'
Z " hex-digit [hex-digit] ..."

R411 hex-digit is digit
    or A
    or B
    or C
    or D
    or E
    or F

R412 signed-real-literal-constant is [sign] real-literal-constant

R413 real-literal-constant is significand [exponent-letter exponent] [_kind-param]
    or digit-string exponent-letter exponent [_kind-param]

R414 significand is digit-string . [digit-string]
    or . digit-string

R415 exponent-letter is E
    or D

R416 exponent is signed-digit-string

Constraint: If both kind-param and exponent-letter are present, exponent-letter must be E.

Constraint: The value of kind-param must specify an approximation method that exists on the processor.

R417 complex-literal-constant is (real-part imag-part)

R418 real-part is signed-int-literal-constant
    or signed-real-literal-constant

R419 imag-part is signed-int-literal-constant
    or signed-real-literal-constant

R420 char-literal-constant is [kind-param_] ' [rep-char] ...'
    or [kind-param_] " [rep-char] ..."

Constraint: The value of kind-param must specify a representation method that exists on the processor.

R421 logical-literal-constant is TRUE. [_kind-param]
    or FALSE. [_kind-param]

Constraint: The value of kind-param must specify a representation method that exists on the processor.

R422 derived-type-def is derived-type-stmt
    [private-sequence-stmt] ...
    component-def-stmt
    [component-def-stmt] ...
    end-type-stmt

R423 private-sequence-stmt is PRIVATE
    or SEQUENCE

R424 derived-type-stmt is TYPE [[, access-spec] ::] type-name

Constraint: The same private-sequence-stmt must not appear more than once in a given derived-type-def.

Constraint: If SEQUENCE is present, all derived types specified in component definitions must be sequence types.

Constraint: An access-spec (5.1.2.2) or a PRIVATE statement within the definition is permitted only if the type definition is within the specification part of a module.

Constraint: If a component of a derived type is of a type declared to be private, either the derived type definition must contain the PRIVATE statement or the derived type must be private.

Constraint: A derived type type-name must not be the same as the name of any intrinsic type nor the same as any other accessible derived type type-name.

R425 end-type-stmt is END TYPE [type-name]

Constraint: If END TYPE is followed by a type-name, the type-name must be the same as that in the corresponding derived-type-stmt.

R426 component-def-stmt is type-spec [[component-attr-spec-list] ::]
    component-decl-list

R427 component-attr-spec is POINTER
    or DIMENSION (component-array-spec)

Constraint: No component-attr-spec may appear more than once in a given component-def-stmt.

Constraint: If the POINTER attribute is not specified for a component, a type-spec in the component-def-stmt must specify an intrinsic type or a previously defined derived type.

Constraint: If the POINTER attribute is specified for a component, a type-spec in the component-def-stmt must specify an intrinsic type or any accessible derived type including the type being defined.

R428 component-array-spec is explicit-shape-spec-list
    or deferred-shape-spec-list

R429 component-decl is component-name [(component-array-spec)]
    [* char-length]

Constraint: If the POINTER attribute is not specified, each component-array-spec must be an explicit-shape-spec-list.

Constraint: If the POINTER attribute is specified, each component-array-spec must be a deferred-shape-spec-list.

Constraint: The char-length option is permitted only if the type specified is character.

Constraint: A char-length in a component-decl must be a constant specification expression (7.1.6.2).

Constraint: Each bound in the explicit-shape-spec (R428) must be a constant specification expression (7.1.6.2).

R430 structure-constructor is type-name (expr-list)

R431 array-constructor is (/ ac-value-list /)

R432 ac-value is expr
    or ac-implied-do

R433 ac-implied-do is (ac-value-list, ac-implied-do-control)

R434 ac-implied-do-control is ac-do-variable = scalar-int-expr,
    scalar-int-expr [scalar-int-expr]

R435 ac-do-variable is scalar-int-variable

Constraint: ac-do-variable must be a named variable.

Constraint: Each ac-value expression in the array-constructor must have the same type and type parameters.

D.1.5 Data object declarations and specifications

R501 type-declaration-stmt is type-spec [[attr-spec] ... ::] entity-decl-list

R502 type-spec is INTEGER [kind-selector]
    or REAL [kind-selector]
    or DOUBLE PRECISION
    or COMPLEX [kind-selector]
    or CHARACTER [char-selector]
    or LOGICAL [kind-selector]
    or TYPE (type-name)

R503 attr-spec is PARAMETER
    or access-spec
    or ALLOCATABLE
    or DIMENSION (array-spec)
    or EXTERNAL
    or INTENT (intent-spec)
    or INTRINSIC
    or OPTIONAL
    or POINTER
    or SAVE
    or TARGET

R504 entity-decl is object-name [(array-spec)]
    [* char-length] [= initialization-expr]
    or function-name [(array-spec)]
    [* char-length]

R505 kind-selector is ([KIND =] scalar-int-initialization-expr)

Constraint: The same attr-spec must not appear more than once in a given type-declaration-stmt.

Constraint: The function-name must be the name of an external function, an intrinsic function, a function dummy procedure, or a statement function.

Constraint: The = initialization-expr must appear if the statement contains a PARAMETER attribute (5.1.2.1).

Constraint: If = initialization-expr appears, a double colon separator must appear before the entity-decl-list.

Constraint: The = initialization-expr must not appear if object-name is a dummy argument, a function result, an object in a named common block unless the type declaration is in a block data program unit, an object in blank common, an allocatable array, a pointer, an external name, an intrinsic name, or an automatic object.

Constraint: The char-length option is permitted only if the type specified is character.

Constraint: The ALLOCATABLE attribute may be used only when declaring an array that is not a dummy argument or a function result.

Constraint: An array declared with a POINTER or an ALLOCATABLE attribute must be specified with an array-spec that is a deferred-shape-spec-list (5.1.2.4.3).

Constraint: An array-spec for a function-name that does not have the POINTER attribute must be an explicit-shape-spec-list.

Constraint: An array-spec for a function-name that does have the POINTER attribute must be a deferred-shape-spec-list.

Constraint: If the POINTER attribute is specified, the TARGET, INTENT, EXTERNAL, or INTRINSIC attribute must not be specified.

Constraint: If the TARGET attribute is specified, the POINTER, EXTERNAL, INTRINSIC, or PARAMETER attribute must not be specified.

Constraint: The PARAMETER attribute must not be specified for dummy arguments, pointers, allocatable arrays, functions, or objects in a common block.

Constraint: The INTENT and OPTIONAL attributes may be specified only for dummy arguments.

Constraint: An entity must not have the PUBLIC attribute if its type has the PRIVATE attribute.

Constraint: The SAVE attribute must not be specified for an object that is in a common block, a dummy argument, a procedure, a function result, or an automatic data object.

Constraint: An entity must not have the EXTERNAL attribute if it has the INTRINSIC attribute.

Constraint: An entity in a type-declaration-stmt must not have the EXTERNAL or INTRINSIC attribute specified unless it is a function.

Constraint: An array must not have both the ALLOCATABLE attribute and the POINTER attribute.

Constraint: An entity must not be given explicitly any attribute more than once in a scoping unit.

Constraint: The value of scalar-int-initialization-expr must be nonnegative and must specify a representation method that exists on the processor.

R506 char-selector is length-selector
    or (LEN = type-param-value,
    KIND = scalar-int-initialization-expr)
    or (type-param-value ,
    [KIND =] scalar-int-initialization-expr)
    or (KIND = scalar-int-initialization-expr
    [, LEN = typeparam-value])

R507 length-selector is ([LEN =] type-param-value)
    or * char-length [,]

R508 char-length is (type-param-value)
    or scalar-int-literal-constant

Constraint: The optional comma in a length-selector is permitted only in a type-spec in a type-declaration-stmt.

Constraint: The optional comma in a length-selector is permitted only if no double colon separator appears in the type-declaration-stmt.

Constraint: The value of scalar-int-initialization-expr must be nonnegative and must specify a representation method that exists on the processor.

Constraint: The scalar-int-literal-constant must not include a kind-param.

R509 type-param-value is specification-expr
or *

Constraint: A function name must not be declared with an asterisk type-param-value if the function is an internal or module function, array-valued, pointer-valued, or recursive.

R510 access-spec is PUBLIC
    or PRIVATE

Constraint: An access-spec attribute may appear only in the scoping unit of a module.

R511 intent-spec is IN
    or OUT
    or INOUT

Constraint: The INTENT attribute must hot be specified for a dummy argument that is a dummy procedure or a dummy pointer.

R512 array-spec is explicit-shape-spec-list
    or assumed-shape-spec-list
    or deferred-shape-spec-list
    or assumed-size-spec

Constraint: The maximum rank is seven.

R513 explicit-shape-spec is [lower-bound:] upper-bound

R514 lower-bound is specification-expr

R515 upper-bound is specification-expr

Constraint: An explicit-shape array whose bounds depend on the values of nonconstant expressions must be a dummy argument, a function result, or an automatic array of a procedure.

R516 assumed-shape-spec is [lower-bound] :

R517 deferred-shape-spec is :

R518 assumed-size-spec is [explicit-shape-spec-list ,] [lower-bound :]

Constraint: The function name of an array-valued function must not be declared as an assumed-size array.

R519 intent-stmt is INTENT (intent-spec) [::] dummy-arg-name-list

Constraint: An intent-stmt may appear only in the specification-part of a subprogram or an interface body (12.3.2.1).

Constraint: dummy-arg-name must not be the name of a dummy procedure or a dummy pointer.

R520 optional-stmt is OPTIONAL [::] dummy-arg-name-list

Constraint: An optional-stmt may occur only in the scoping unit of a subprogram or an interface body.

R521 access-stmt is access-spec [[::] access-id-list]

R522 access-id is use-name
    or generic-spec

Constraint: An access-stmt may appear only in the scoping unit of a module. Only one accessibility statement with an omitted access-id-list is permitted in the scoping unit of a module.

Constraint: Each use-name must be the name of a named variable, procedure, derived type, named constant, or namelist group.

Constraint: A module procedure that has a dummy argument or function result of a type that has PRIVATE accessibility must have PRIVATE accessibility and must not have a generic identifier that has PUBLIC accessibility.

R523 save-stmt is SAVE [[::] saved-entity-list]

R524 saved-entity is object-name
    or / common-block-name

Constraint: An object-name must not be a dummy argument name, a procedure name, a function result name, an automatic data object name, or the name of an entity in a common block.

Constraint: If a SAVE statement with an omitted saved entity list occurs in a scoping unit, no other explicit occurrence of the SAVE attribute or SAVE statement is permitted in the same scoping unit.

R525 dimension-stmt is DIMENSION [::] array-name (array-spec)
    [, array-name (array-spec)] ...

R526 allocatable-stmt is ALLOCATABLE [::] array-name
    [(deferred-shape-spec-list)]
    [, array-name
    [(deferred-shape-spec-list)]] ...

Constraint: The array-name must not be a dummy argument or function result.

Constraint: If the DIMENSION attribute for an array-name is specified elsewhere in the scoping unit, the array-spec must be a deferred-shape-spec-list.

R527 pointer-stmt is POINTER [::] object-name
    [(deferred-shape-spec-list)]
    [, object-name
    [(deferred-shape-spec-list)]] ...

Constraint: The INTENT attribute must not be specified for an object-name.

Constraint: If the DIMENSION attribute for an object-name is specified elsewhere in the scoping unit, the array-spec must be a deferred-shape-spec-list.

Constraint: The PARAMETER attribute must not be specified for an object-name.

R528 target-stmt is TARGET [::] object-name [(array-spec)]
    [, object-name [(array-spec)]] ...

Constraint: The PARAMETER attribute must not be specified for an object-name.

R529 data-stmt is DATA data-stmt-set [[,] data-stmt-set] ...

R530 data-stmt-set is data-stmt-object-list / data-stmt-value-list

R531 data-stmt-object is variable
    or data-implied-do

R532 data-stmt-value is [data-stmt-repeat *] data-stmt-constant

R533 data-stmt-constant is scalar-constant
    or signed-in t-literal-constant
    or signed-real-literal-constant
    or structure-constructor
    or boz-literal-constant

R534 data-stmt-repeat is scalar-int-constant

R535 data-implied-do is (data-i-do-object-list , data-i-do-variable =
    scalar-int-expr, scalar-int-expr
    [, scalar-int-expr])

R536 data-i-do-object is array-element
    or scalar-structure-component
    or data-implied-do

Constraint: The array-element must not have a constant parent.

Constraint: The scalar-structure-component must not have a constant parent.

R537 data-i-do-variable is scalar-int-variable

Constraint: data-i-do-variable must be a named variable.

Constraint: The DATA statement repeat factor must be positive or zero. If the DATA statement repeat factor is a named constant, it must have been declared previously in the scoping unit or made accessible by use association or host association.

Constraint: If a data-stmt-constant is a structure-constructor, each component must be an initialization expression.

Constraint: In a variable that is a data-stmt-object, any subscript, section subscript, substring starting point, and substring ending point must be an initialization expression.

Constraint: A variable whose name or designator is included in a data-stmt-object-list or a data-i-do-object-list must not be: a dummy argument, made accessible by use association or host association, in a named common block unless the DATA statement is in a block data program unit, in a blank common block, a function name, a function result name, an automatic object, a pointer, or an allocatable array.

Constraint: In an array-element or a scalar-structure-component that is a data-i-do-object, any subscript must be an expression whose primaries are either constants or DO variables of the containing data-implied-dos, and each operation must be intrinsic.

Constraint: A scalar-int-expr of a data-implied-do must involve as primaries only constants or DO variables of the containing data-implied-dos, and each operation must be intrinsic.

R538 parameter-stmt is PARAMETER (named-constant-def-list)

R539 named-constant-def is named-constant = initialization-expr

R540 implicit-stmt is IMPLICIT implicit-spec-list
    or IMPLICIT NONE

R541 implicit-spec is type-spec (letter-spec-list)

R542 letter-spec is letter [- letter]

Constraint: If IMPLICIT NONE is specified in a scoping unit, it must precede any PARAMETER statements that appear in the scoping unit and there must be no other IMPLICIT statements in the scoping unit.

Constraint: If the minus and second letter appear, the second letter must follow the first letter alphabetically.

R543 namelist-stmt is NAMELIST / namelist-group-name /
    namelist-group-object-list
    [[,] / namelist-group-name /
    namelist-group-object-list] ...

R544 namelist-group-object is variable-name

Constraint: A namelist-group-object must not be an array dummy argument with a nonconstant bound, a variable with nonconstant character length, an automatic object, a pointer, a variable of a type that has an ultimate component that is a pointer, or an allocatable array.

Constraint: If a namelist-group-name has the PUBLIC attribute, no item in the namelist-group-object-list may have the PRIVATE attribute.

R545 equivalence-stmt is EQUIVALENCE equivalence-set-list

R546 equivalence-set is (equivalence-object, equivalence-object-list)

R547 equivalence-object is variable-name
    or array-element
    or substring

Constraint: An equivalence-object must not be a dummy argument, a pointer, an allocatable array, an object of a nonsequence derived type or of a sequence derived type containing a pointer at any level of component selection, an automatic object, a function name, an entry name, a result name, a named constant, a structure component, or a subobject of any of the preceding objects.

Constraint: Each subscript or substring range expression in an equivalence-object must be an integer initialization expression (7.1.6.1).

Constraint: If an equivalence-object is of type default integer, default real, double precision real, default complex, default logical, or numeric sequence type, all of the objects in the equivalence set must be of these types.

Constraint: If an equivalence-object is of type default character or character sequence type, all of the objects in the equivalence set must be of these types.

Constraint: If an equivalence-object is of a derived type that is not a numeric sequence or

character sequence type, all of the objects in the equivalence set must be of the same type.
Constraint: If an equivalence-object is of an intrinsic type other than default integer, default real, double precision real, default complex, default logical, or default character, all of the objects in the equivalence set must be of the same type with the same kind type parameter value.

R548 common-stmt is COMMON
[/ [common-block-name] /]
common-block-object-list
[[,] / [common-block-name] /
common-block-object-list] ...

R549 common-block-object is variable-name [(explicit-shape-spec-list)]

Constraint: Only one appearance of a given variable-name is permitted in all common-block-object-lists within a scoping unit.

Constraint: A common-block-object must not be a dummy argument, an allocatable array, an automatic object, a function name, an entry name, or a result name.

Constraint: Each bound in the explicit-shape-spec must be a constant specification expression (7.1.6.2).

Constraint: If a common-block-object is of a derived type, it must be a sequence type (4.4.1).

Constraint: If a variable-name appears with an explicit-shape-spec-list, it must not have the POINTER attribute.

D.1.6 Use of data objects

R601 variable is scalar-variable-name
or array-variable-name
or subobject

Constraint: array-variable-name must be the name of a data object that is an array.

Constraint: array-variable-name must not have the PARAMETER attribute.

Constraint: scalar-variable-name must not have the PARAMETER attribute.

Constraint: subobject must not be a subobject designator (for example, a substring) whose parent is a constant.

R602 subobject is array-element
or array-section
or structure-component
or substring

R603 logical-variable is variable
Constraint: logical-variable must be of type logical.

R604 default-logical-variable is variable
Constraint: default-logical-variable must be of type default logical.

R605 char-variable is variable
Constraint: char-variable must be of type character.

R606 default-char-variable is variable
Constraint: default-char-variable must be of type default character.

R607 int-variable is variable
Constraint: int-variable must be of type integer.

R608 default-int-variable is variable
Constraint: default-int-variable must be of type default integer.

R609 substring is parent-string (substring-range

R610 parent-string is scalar-variable-name
or array-element
or scalar-structure-component
or scalar-constant

R611 substring-range is [scalar-int-expr] : [scalar-int-expr]
Constraint: parent-string must be of type character.

R612 data-ref is part-ref [% part-ref] ...

R613 part-ref is part-name [(section-subscript-list)]]
Constraint: In a data-ref, each part-name except the rightmost must be of derived type.
Constraint: In a data-ref, each part-name except the leftmost must be the name of a

component of the derived type definition of the type of the preceding part-name.
Constraint: In a part-ref containing a section-subscript-list, the number of section-subscripts must equal the rank of part-name.
Constraint: In a data-ref, there must not be more than one part-ref with nonzero rank. A part-name to the right of a part-ref with nonzero rank must not have the POINTER attribute.
R614 structure-component is data-ref
Constraint: In a structure-component, there must be more than one part-ref and the rightmost part-ref must be of the form part-name.
R6J5 array-element is data-ref
Constraint: In an array-element, every part-ref must have rank zero and the last part-ref must contain a subscript-list.
R616 array-section is data-ref [(substring-range)]
Constraint: In an array-section, exactly one part-ref must have nonzero rank, and either the final part-ref has a section-subscript-list with nonzero rank or another part-ref has nonzero rank.
Constraint: In an array-section with a substring-range, the rightmost part-name must be of type character.
R617 subscript is scalar-int-expr
R618 section-subscript is subscript
    or subscript-triplet
    or vector-subscript
R619 subscript-triplet is [subscript] : [subscript]
    [stride]
R620 stride is scalar-int-expr
R621 vector-subscript is int-expr
Constraint: A vector-subscript must be an integer array expression of rank one.
Constraint: The second subscript must not be omitted from a subscript-triplet in the last dimension of an assumed-size array.

R622 allocate-stmt is ALLOCATE
(allocation-list
    [, STAT = stat-variable])
R623 stat-variable is scalar-int-variable
R624 allocation is allocate-object [(allocate-shape-spec-list)]
R625 allocate-object is variable-name
    or structure-component
R626 allocate-shape-spec is [allocate-lower-bound :] allocate-upper-bound
R627 allocate-lower-bound is scalar-int-expr
R628 allocate-upper-bound is scalar-int-expr
Constraint: Each allocate-object must be a pointer or an allocatable array.
Constraint: The number of allocate-shape-specs in an allocate-shape-spec-list must be the same as the rank of the pointer or allocatable array.
R629 nullify-stmt is NULLIFY (pointer-object-list
R630 pointer-object is variable-name
    or structure-component
Constraint: Each pointer-object must have the POINTER attribute.
R631 deallocate-stmt is DEALLOCATE (allocate-object-list
    [, STAT = stat-variable])
Constraint: Each allocate-object must be a pointer or an allocatable array.

D.1.7 Expressions and assignment

R701 primary is constant
    or constant-subobject
    or variable
    or array-constructor
    or structure-constructor
    or function-reference
    or (expr)
R702 constant-subobject is subobject
Constraint: subobject must be a subobject designator whose parent is a constant.
Constraint: A variable that is a primary must not be an assumed-size array.

R703 level-1-expr is [defined-unary-op] primary
R704 defined-unary-op is . letter [letter] ... .
Constraint: A defined-in nary-op must not contain more than 31 letters and must not be the same as any intrinsic-operator or logical-literal-constant.
R705 mult-operand is level-1-expr [power-op mult-operand]
R706 add-operand is [add-operand mult-op] mult-operand
R707 level-2-expr is [[level-2-expr] add-op] add-operand
R708 power-op is **
R700 mult-op is *
 or /
R710 add-op is +
 or -
R711 level-3-expr is [level-3-expr concat-op] level-2-expr
R712 concat-op is //
R713 level-4-expr is [level-3-expr rel-op] level-3-expr
R714 rel-op is .EQ.
 or NE.
 or .LT.
 or .LE.
 or .GT.
 or .GE.
 or = =
 or / =or
 or <
 or < =
 or >
 or > =
R715 and-operand is [not-op] level-4-expr
R716 or-operand is [or-operand and-op] and-operand
R717 equiv-operand is [equiv-operand or-op] or-operand
R718 level-5-expr is [level-5-expr equiv-op] equiv-operand

R719 not-op is .NOT.
R720 and-op is .AND.
R721 or-op is OR.
R722 equiv-op is .EQV.
 or .NEQV.
R723 expr is [expr defined-binary-op] level-5-expr
R724 defined-binary-op is letter[letter]
Constraint: A defined-binary-op must not contain more than 31 letters and must not be the same as any intrinsic-operator or logical-literal-constant.
R725 logical-expr is expr
Constraint: logical-expr must be type logical.
R726 char-expr is expr
Constraint: char-expr must be type character.
R727 default-char-expr is expr
Constraint: default-char-expr must be of type default character.
R728 int-expr is expr
Constraint: int-expr must be type integer.
R729 numeric-expr is expr
Constraint: numeric-expr must be of type integer, real or complex.
R730 initialization-expr is expr
Constraint: An initialization-expr must be an initialization expression.
R731 char-initialization-expr is char-expr
Constraint: A char-initialization-expr must be an initialization expression.
R732 int-initialization-expr is int-expr
Constraint: An int-initialization-expr must be an initialization expression.
R733 logical-initialization-expr is logical-expr
Constraint: A logical-initialization-expr must be an initialization expression.
R734 specification-expr is scalar-int-expr
Constraint: The scalar-int-expr must be a restricted expression.
R735 assignment-stmt is variable = expr
Constraint: A variable in an assignment-stmt must not be an assumed-size array.

R736 pointer-assignment-stmt is pointer-object = target
R737 target is variable
or expr
Constraint: The pointer-object must have the POINTER attribute.
Constraint: The variable must have the TARGET attribute or be a subobject of an object with the TARGET attribute, or it must have the POINTER attribute.
Constraint: The target must be of the same type, type parameters, and rank as the pointer.
Constraint: The target must not be an array section with a vector subscript.
Constraint: The expr must deliver a pointer result.
R738 where-stmt is WHERE (mask-expr) assignment-stmt
R739 where-construct is where-construct-stmt
    [assignment-stmt] ...
  [elsewhere-stmt
    [assignment-stmt] ...]
  end-where-stmt
R740 where-construct-stmt is WHERE (mask-expr)
R741 mask-expr is logical-expr
R742 elsewhere-stmt is ELSEWHERE
R743 end-where-stmt is END WHERE
Constraint: In each assignment-stmt, the mask-expr and the variable being defined must be arrays of the same shape.
Constraint: The assignment-stmt must not be a defined assignment.

D.1.8 Execution control

R801 block is [execution-part-construct] ...
R802 if-construct is if-then-stmt
    block
  [else-if-stmt
    block] ...
  [else-stmt
    block]
  end-if-stmt
R803 if-then-stmt is [if-construct-name :] IF (scalar-logical-expr) THEN
R804 else-if-stmt is ELSE IF (scalar-logical-expr) THEN [if-construct-name]
R805 else-stmt is ELSE [if-construct-name]
R806 end-if-stmt is END IF [if-construct-name]
Constraint: If the if-then-stmt of an if-construct is identified by an if-construct-name, the corresponding end-if-stmt must specify the same if-construct-name. If the if-then-stmt of an if-construct is not identified by an if-construct-name, the corresponding end-if-stmt must not specify an if- construct-name. If an else-if-stmt or else-stmt is identified by an if-construct-name, the corresponding if-then-stmt must specify the same if-construct-name.
R807 if-stmt is IF (scalarlogical-expr) action-stmt
Constraint: The action-stmt in the if-stmt must not be an if-stmt, end-program-stmt, end-function-stmt, or end-subroutine-stmt.
R808 case-construct is select-case-stmt
  [case-stmt
  block] ...
  end-select-stmt
R809 select-case-stmt is [case-construct-name :] SELECT CASE (case-expr)
R810 case-stmt is CASE case-selector [case-construct-name]
R811 end-select-stmt is END SELECT [case-construct-name]
Constraint: If the select-case-stmt of a case-construct is identified by a case-construct-name, the corresponding end-select-stmt must specify the same case-construct-name. If the select-case-stmt of a case-construct is not identified by a case-construct-name, the corresponding end-select-stmt must not specify a case-construct-name. If a case-stmt is identified by a case-construct-name, the corresponding se-

lect-case-stmt must specify the same case-construct-name.

R812 case-expr is scalar-int-expr
    or scalar-char-expr
    or scalar-logical-expr

R813 case-selector is (case-value-range-list)
    or DEFAULT

Constraint: No more than one of the selectors of one of the CASE statements may be DEFAULT.

R814 case-value-range is case-value
    or case-value :
    or : case-value
    or case-value : case-value

R815 case-value is scalar-int-initialization-expr
    or scalar-char-initialization-expr
    or scalar-logical-initialization-expr

Constraint: For a given case-construct, each case-value must be of the same type as case-expr. For character type, length differences are allowed, but the kind type parameters must be the same.

Constraint: A case-value-range using a colon must not be used if case-expr is of type logical.

Constraint: For a given case-construct, the case-value-ranges must not overlap; that is, there must be no possible value of the case-expr that matches more than one case-value-range.

R816 do-construct is block-do-construct
    or nonblock-do-construct

R817 block-do-construct is do-stmt
      do-block
      end-do

R818 do-stmt is label-do-stmt
    or nonlabel-do-stmt

R819 label-do-stmt is [do-construct-name :] DO label [loop-control]

R820 nonlabel-do-stmt is [do-construct-name :] DO [loop-control]

R821 loop-control is [,] do-variable scalar-numeric-expr, scalar-numeric-expr
[, scalar-numeric-expr]
or [,] WHILE (scalar-logical-expr)

R822 do-variable is scalar-variable

Constraint: The do-variable must be a named scalar variable of type integer, default real, or double precision real.

Constraint: Each scalar-numeric-expr in loop-control must be of type integer, default real, or double precision real.

R823 do-block is block

R824 end-do is end-do-stmt
    or continue-stmt

R825 end-do-stmt is END DO [do-construct-name]

Constraint: If the do-stmt of a block-do-construct is identified by a do-construct-name, the corresponding end-do must be an end-do-stmt specifying the same do-construct-name. If the do-stmt of a block-do-construct is not identified by a do-construct-name, the corresponding end-do must not specify a do-construct-name.

Constraint: If the do-stmt is a nonlabel-do-stmt, the corresponding end-do must be an end-do-stmt.

Constraint: If the do-stmt is a label-do-stmt, the corresponding end-do must be identified with the same label.

R826 nonblock-do-construct is action-term-do-construct
    or outer-shared-do-construct

R827 action-term-do-construct is label-do-stmt
      do-body
      do-term-action-stmt

R828 do-body is [execution-part-construct]

R829 do-term-action-stmt is action-stmt

Constraint: A do-term-action-stmt must not be a continue-stmt, a goto-stmt, a return-stmt, a stop-stmt, an exit-stmt, a cycle-stmt, an end-function-stmt, an end-subroutine-stmt, an

end-program-stmt, an arithmetic-if-stmt, or an assigned-goto-stmt.
Constraint: The do-term-action-stmt must be identified with a label and the corresponding label-do-stmt must refer to the same label.
R830 outer-shared-do-construct is label-do-stmt
    do-body
    shared-term-do-construct
R831 shared-term-do-construct is outer-shared-do-construct
    or inner-shared-do-construct
R832 inner-shared-do-construct is label-do-stmt
    do-body
    do-term-shared-stmt
R833 do-term-shared-stmt is action-stmt
Constraint: A do-term-shared-stmt must not be a goto-stmt, a return-stmt, a stop-stmt, an exit-stmt, a cycle-stmt, an end-function-stmt, an end-subroutine-stmt, an end-program-stmt, an arithmetic-if-stmt, or an assigned-goto-stmt.
Constraint: The do-ter,n-shared-stmt must be identified with a label and all of the label-do-stmts of the shared-term-do-construct must refer to the same label.
R834 cycle-stmt is CYCLE [do-construct-name]
Constraint: If a cycle-stmt refers to a do-construct-name, it must be within the range of that do-construct; otherwise, it must be within the range of at least one do-construct
R835 exit-stmt is EXIT [do-construct-name]
Constraint: If an exit-stmt refers to a do-construct-name, it must be within the range of that do-
construct; otherwise, it must be within the range of at least one do-construct.
R836 go to-stmt is GO TO label
Constraint: The label must be the statement label of a branch target statement that appears in the same scoping unit as the goto-stmt.

R837 computed-goto-stmt is GO TO (label-list) [,] scalar-int-expr
Constraint: Each label in label-list must be the statement label of a branch target statement that appears in the same scoping unit as the computed-go to-s tint.
R838 assign-stmt is ASSIGN label TO scalar-int-variable
Constraint: The label must be the statement label of a branch target statement or format-stmt that appears in the same scoping unit as the assign-stmt.
Constraint: scalar-int-variable must be named and of type default integer.
R839 assigned-goto-stmt is GO TO scalar-int-variable [[,] (label-list)]
Constraint: Each label in label-list must be the statement label of a branch target statement that appears in the same scoping
    unit as the assigned-goto-stmt.
Constraint: scalar-int-variable must be named and of type default integer.
R840 arithmetic-if-stmt is IF (scalar-numeric-expr) label , label , label
Constraint: Each label must be the label of a branch target statement that appears in the same scoping unit as the arithmetic-if-
Constraint: The scalar-numeric-expr must not be of type complex.
R841 continue-stmt is CONTINUE
R842 stop-stmt is STOP [stop-code]
R843 stop-code is scalar-char-constant
    or digit [digit [digit [digit [digit]]]]
Constraint: scalar-char-constant must be of type default character.
R844 pause-stmt is PAUSE [stop-code]

D.1.9 input/output statements

R901 io-unit is external-file-unit
    or *
    or internal-file-unit
R902 external-file-unit is scalar-int-expr

R903 internal-file-unit is default-char-variable
Constraint: The default-char-variable must not be an array section with a vector subscript.
R904 open-stmt is OPEN (connect-spec-list)
R905 connect-spec is [UNIT =] external-file-unit
    or IOSTAT = scalar-default-int-variable
    or ERR = label
    or FILE = file-name-expr
    or STATUS scalardefault-char-expr
    or ACCESS = scalar-default-char-expr
    or FORM = scalar-default-char-expr
    or RECL = scalar-int-expr
    or BLANK = scalar-default-char-expr
    or POSITION = scalar-default-char-expr
    or ACTION = scalar-default-charexpr
    or DELIM = scalardefault-char-expr
    or PAD = scalardefaultchar-expr
R906 file-name-expr is scalar-default-char-expr
Constraint: If the optional characters UNIT = are omitted from the unit specifier, the unit specifier must be the first item in the connect-spec-list.
Constraint: Each specifier must not appear more than once in a given open-stmt; an external-file-unit
must be specified.
Constraint: The label used in the ERR = specifier must be the statement label of a branch target statement that appears in the same scoping unit as the OPEN statement.
R907 close-stmt is CLOSE (close-spec-list)
R908 close-spec is [UNIT =] external-file-unit
    or IOSTAT scalar-int-default-variable
    or ERR = label
    or STATUS = scalardefault-char-expr
Constraint: If the optional characters UNIT = are omitted from the unit specifier, the unit specifier must be the first item in the close-spec-list.

Constraint: Each specifier must not appear more than once in a given close-stmt; an external-file-unit must be specified.
Constraint: The label used in the ERR = specifier must be the statement label of a branch target statement that appears in the same scoping unit as the CLOSE statement.
R909 read-stmt is READ (iocontrol-spec-list) [input-item-list]
    or READ format [, input-item-list]
R910 write-stmt is WRITE (iocontrol-spec-list) [output-item-list]
R911 print-stmt is PRINT format [, outputitem-list]
R912 io-control-spec is [UNIT =] io-unit
    or [FMT =] format
    or [NML =] namelist-group-name
    or REC = scalar-int-expr
    or IOSTAT = scalardefault-int-variable
    or ERR = label
    or END = label
    or ADVANCE = scalardefault-char-expr
    or SIZE = scalardefault-int-variable
    or EOR = label
Constraint: An iocontrol-spec-list must contain exactly one io-unit and may contain at most one of each of the other specifiers.
Constraint: An END, EOR =, or SIZE = specifier must not appear in a write-stmt.
Constraint: The label in the ERR =, EOR =, or END = specifier must be the statement label of a branch target statement that appears in the same scoping unit as the data transfer statement.
Constraint: A namelistgroup-name must not be present if an input-item-list or an output-item-list is present in the data transfer statement.
Constraint: An iocontrolspec-list must not contain both a format and a namelistgroup-name.
Constraint: If the optional characters UNIT = are omitted from the unit specifier, the unit specifier must be the first item in the control information list.

Constraint: If the optional characters FMT = are omitted from the format specifier, the format specifier must be the second item in the control information list and the first item must be the unit specifier without the optional characters UNIT

Constraint: If the optional characters NML= are omitted from the namelist specifier, the namelist specifier must be the second item in the control information list and the first item must be the unit specifier without the optional characters UNIT.

Constraint: If the unit specifier specifies an internal file, the io-control-spec-list must not contain a REC = specifier or a namelist-group-name.

Constraint: If the REC = specifier is present, an END = specifier must not appear, a namelist-group-name must not appear, and the format, if any, must not be an asterisk specifying list-directed input/output.

Constraint: An ADVANCE = specifier may be present only in a formatted sequential input/output statement with explicit format specification (10.1) whose control information list does not contain an internal file unit specifier.

Constraint: If an EOR = specifier is present, an ADVANCE = specifier also must appear.

R913 format is defaultcharexpr
   or label
   or
   or scalar-default-int-variable

Constraint: The label must be the label of a FORMAT statement that appears in the same scoping unit as the statement containing the format specifier.

R914 input-item is variable
   or io-implied-do

R915 output-item is expr
   or io-implied-do

R916 io-implied-do is (io-implied-do-object-list, io-implied-do-control)

R917 io-implied-do-object is input-item
   or output-item

R918 io-implied-do-control is do-variable = scalar-numeric-expr,
       scalar-numeric-expr [, scalar-numeric-expr]

Constraint: A variable that is an input-item must not be an assumed-size array.

Constraint: The do-variable must be a scalar of type integer, default real. or double precision real.

Constraint: Each scalar-numeric-expr in an io-implied-do-control must be of type integer, default real. or double precision real.

Constraint: In an input-item-list, an io-implied-do-object must be an input-item. In an output-item-list, an io-implied-do-object must be an output-item.

R919 backspace-stmt is BACKSPACE external-file-unit
   or BACKSPACE (position-spec-list)

R920 endfile-stmt is ENDFILE external-file-unit
   or ENDFILE (position-spec-list)

R921 rewind-stmt is REWIND external-file-unit
   or REWIND (position-spec-list)

R922 position-spec is [UNIT =] external-file-unit
   or IOSTAT = scalar-default-int-variable
   or ERR = label

Constraint: The label in the ERR = specifier must be the statement label of a branch target statement that appears in the same scoping unit as the file positioning statement.

Constraint: If the optional characters UNIT = are omitted from the unit specifier, the unit specifier must be the first item in the position-spec-list.

Constraint: A position-spec-list must contain exactly one external-file-unit and may contain at most one of each of the other specifiers.

R923 inquire-stmt is INQUIRE
(inquire-spec-list
or INQUIRE (IOLENGTH = scalar-default-int-variable)
    output-item-list
R924 inquire-spec is [UNIT =] external-file-unit
    or FILE = file-name-expr
    or IOSTAT = scalar-default-int-variable
    or ERR = label
    or EXIST = scalar-default-logical-variable
    or OPENED = scalar-default-logical-variable
    or NUMBER = scalar-default-int-variable
    or NAMED = scalar-default-logical-variable
    or NAME = scalar-default-char-variable
    or ACCESS = scalar-default-char-variable
    or SEQUENTIAL =
        scalar-default-char-variable
    or DIRECT = scalar-default-char-variable
    or FORM = scalar-default-char-variable
    or FORMATTED =
        scalar-default-char-variable
    or UNFORMATTED =
        scalar-default-char-variable
    or RECL = scalar- default- in t-variable
    or NEXTREC = scalar-default-int-variable
    or BLANK = scalar-default-char-variable
    or POSITION = scalar-default-char-variable
    or ACTION = scalar-default-char-variable
    or READ = scalar-default-char-variable
    or WRITE = scalar-default-char-variable
    or READWRITE =
        scalar-default-char-variable
    or DELIM = scalar-default-char-variable
    or PAD = scalar-default-char-variable

Constraint: An inquire-spec-list must contain one FILE = specifier or one UNIT = specifier, but not both, and at most one of each of the other specifiers.

Constraint: In the inquire by unit form of the INQUIRE statement, if the optional characters UNIT = are omitted from the unit specifier, the unit specifier must be the first item in the inquire-spec-list.

D.1.10 Input/output editing

R1001 format-stmt is FORMAT format-specification

R1002 format-specification is ([format-item-list]

Constraint: The format-stmt must be labeled.

Constraint: The comma used to separate format-items in a format-item-list may be omitted as follows:

(1) Between a P edit descriptor and an immediately following F, E, EN, ES, D, or G edit descriptor (10.6.5)

(2) Before a slash edit descriptor when the optional repeat specification is not present (10.6.2)

(3) Alter a slash edit descriptor

(4) Before or after a colon edit descriptor (10.6.3)

R1003 format-item is [r] data-edit-desc
    or control-edit-desc
    or char-string-edit-desc
    or [r] (format-item-list)

R1004 r is int-literal-constant

Constraint: r must be positive.

Constraint: r must not have a kind parameter specified for it.

R1005 data-edit-desc is I w [. m]
    or B w [. m]
    or O w [. m]
    or Z w [. m]
    or F w. d
    or E w. d [E e]
    or EN w. d [E e]
    or ES w . d [E e]
    or G w. d [E e]
    or L w
    or A [w]
    or D w. d

R1006 w is int-literal-constant

R1007 in is int-literal-constant

R1008 d is int-literal-constant
R1009 e is int-literal-constant
Constraint: w and e must be positive.
Constraint: w, in, d, and e must not have kind parameters specified for them.
R1010 control-edit-desc is position-edit-desc
    or [r]
    or :
    or sign-edit-desc
    or k P
    or blank-interp-edit-desc
R1011 k is signed-int-literal-constant
Constraint: k must not have a kind parameter specified for it.
R1012 position-edit-desc is T n
    or TL n
    or TR n
    or n X
R1013 n is int-literal-constant
Constraint: n must be positive.
Constraint: n must not have a kind parameter specified for it.
R1014 sign-edit-desc is S
    or SP
    or SS
R1015 blank-interp-edit-desc is BN
    or BZ
R1016 char-string-edit-desc is char-literal-constant
    or c H rep-char [rep-char] ...
R1017 c is int-literal-constant
Constraint: c must be positive.
Constraint: c must not have a kind parameter specified for it.
Constraint: The rep-char in the cH form must be of default character type.
Constraint: The char-literal-constant must not have a kind parameter specified for it.

D.1.11 Program units

R1101 main-program is [program-stmt]
    [specification-part]
    [execution-part]
    [internal-subprogram-part]
    end-program-stmt
R1102 program-stmt is PROGRAM program-name
R1103 end-program-stmt is END [PROGRAM [program-name]]
Constraint: In a main-program, the execution-part must not contain a RETURN statement or an ENTRY statement.
Constraint: The program-name may be included in the end-program-stmt only if the optional program-stmt is used and, if included, must be identical to the program-name specified in the program-stmt.
Constraint: An automatic object must not appear in the specification-part (R204) of a main program.
R1104 module is module-stmt
    [specification-part]
    [module-subprogram-part]
    end-module-stmt
R1105 module-stmt is MODULE module-name
R1106 end-module-stmt is END [MODULE [module-name]]
Constraint: If the module-name is specified in the end-module-stmt, it must be identical to the module-name specified in the module-stmt.
Constraint: A module specification-part must not contain a stmt-function-stmt, an entry-stmt, or a format-stmt.
Constraint: An automatic object must not appear in the specification-part (R204) of a module.
R1107 use-stmt is USE module-name [, rename-list]
    or USE module-name, ONLY : [only-list]
R1108 rename is local-name => use-name
R1109 only is access-id
    or [local-name =>] use-name
Constraint: Each access-id must be a public entity in the module.

Constraint: Each use-name must be the name of a public entity in the module.

R1110 block-data is block-data-stmt
    [specification-part]
    end-block-data-stmt

R1111 block-data-stmt is BLOCK DATA [block-data-name]

R1112 end-block-data-stmt is END [BLOCK DATA [block-data-name]]

Constraint: The block-data-name may be included in the end-block-data-stmt only if it was provided in the block-data-stmt and, if included, must be identical to the block-data-name in the block-data-stmt.

Constraint: A block-data specification-part may contain only USE statements, type declaration statements, IMPLICIT statements, PARAMETER statements, derived-type definitions, and the following specification statements: COMMON, DATA, DIMENSION, EQUIVALENCE, INTRINSIC, POINTER, SAVE, and TARGET.

Constraint: A type declaration statement in a block-data specification-part must not contain ALLOCATABLE, EXTERNAL, INTENT, OPTIONAL, PRIVATE, or PUBLIC attribute specifiers.

D.1.12 Procedures

R1201 interface-block is interface-stmt
    [interface-body]
    [module-procedure-stmt]
    end-interface-stmt

R1202 interface-stmt is INTERFACE [generic-spec]

R1203 end-interface-stmt is END INTERFACE

R1204 interface-body is function-stmt
      [specification-part]
    end-function-stmt
    or subroutine-stmt
      [specification-part]
    end-subroutine-stmt

R1205 module-procedure-stmt is MODULE PROCEDURE procedure-name-list

R1206 generic-spec is generic-name
    or OPERATOR (defined-operator)
    or ASSIGNMENT (=)

Constraint: An interface-body must not contain an entry-stmt, data-stmt, format-stmt, or stmt-function-stmt.

Constraint: The MODULE PROCEDURE specification is allowed only if the interface-block has a generic-spec and has a host that is a module or accesses a module by use association; each procedure-name must be the name of a module procedure that is accessible in the host.

Constraint: An interface-block must not appear in a BLOCK DATA program unit.

Constraint: An interface-block in a subprogram must not contain an interface-body for a procedure defined by that subprogram.

R1207 external-stmt is EXTERNAL external-name-list

R1208 intrinsic-stmt is INTRINSIC intrinsic-procedure-name-list

Constraint: Each intrinsic-procedure-name must be the name of an intrinsic procedure.

R1209 function-reference is function-name ([actual-arg-spec-list])

Constraint: The actual-arg-spec-list for a function reference must not contain an alt-return-spec.

R1210 call-stmt is CALL subroutine-name [([actual-arg-spec-list])]

R1211 actual-arg-spec is [keyword =] actual-arg

R1212 keyword is dummy-arg-name

R1213 actual-arg is expr
    or variable
    or procedure-name
    or alt-return-spec

R1214 alt-return-spec is * label

Constraint: The keyword = must not appear if the interface of the procedure is implicit in the scoping unit.

Constraint: The keyword = may be omitted from an actual-arg-spec only if the keyword = has been omitted from each preceding actual-arg-spec in the argument list.

Constraint: Each keyword must be the name of a dummy argument in the explicit interface of the procedure.

Constraint: A procedure-name actual-arg must not be the name of an internal procedure or of a statement function and must not be the generic name of a procedure (12.3.2.1, 13.1).

Constraint: The label used in the alt-return-spec must be the statement label of a branch target statement that appears in the same scoping unit as the call-stmt.

R1215 function-subprogram is function-stmt
[specification-part]
[execution-part]
[internal-subprogram-part]
end-function-stmt

R1216 function-stmt is [prefix] FUNCTION function-name
([dummy-arg-name-list])
[RESULT (result-name)]

Constraint: If RESULT is specified, the function'-name must not appear in any specification statement inthe scoping unit of the function subprogram.

R1217 prefix is type-spec [RECURSIVE]
or RECURSIVE [type-spec]

R1218 end-function-stmt is END [FUNCTION [function-name]]

Constraint: If RESULT is specified, result-name must not be the same as function-name.

Constraint: FUNCTION must be present on the end-function-stmt of an internal or module function.

Constraint: An internal function must not contain an ENTRY statement.

Constraint: An internal function must not contain an internal-subprogram-part.

Constraint: If a function-name is present on the end-function-stmt, it must be identical to the function-name specified in the function-stmt.

R1219 subroutine-subprogram is subroutine-stmt
[specification-part]
[execution-part]
[internal-subprogram-part]
end-subroutine-stmt

R1220 subroutine-stmt is [RECURSIVE] SUBROUTINE subroutine-name
[([dummy-arg-list])]

R1221 dummy-arg is dummy-arg-name
or *

R1222 end-subroutine-stmt is END [SUBROUTINE [subroutine-name]]

Constraint: SUBROUTINE must be present on the end-subroutine-stmt of an internal or module subroutine.

Constraint: An internal subroutine must not contain an ENTRY statement.

Constraint: An internal subroutine must not contain an internalsubprogram-part.

Constraint: If a subroutine-name is present on the end-subroutine-stmt, it must be identical to the subroutine-name specified in the subroutine-stmt.

R1223 entry-stmt is ENTRY entry-name [([dummy-arg-list])
[RESULT (result-name)]]

Constraint: If RESULT is specified, the entry-name must not appear in any specification statement in the scoping unit of the function program.

Constraint: An entry-stmt may appear only in an external-subprogram or module-subprogram. An entry-stmt must not appear within an executable-construct.

Constraint: RESULT may be present only if the entry-stmt is contained in a function subprogram.

Constraint: Within the subprogram containing the entry-stmt, the entry-name must not appear as a dummy argument in the FUNCTION or SUBROUTINE statement or in another ENTRY statement and it must not appear in an EXTERNAL or INTRINSIC statement.

Constraint: A dummy-arg may be an alternate return indicator only if the ENTRY statement is contained in a subroutine subprogram.

Constraint: If RESULT is specified, result-name must not be the same as entry-name.

R1224 return-stmt is RETURN [scalar-nt-expr]

Constraint: The return-stmt must be contained in the scoping unit of a function or subroutine subprogram.

Constraint: The scalar-int-expr is allowed only in the scoping unit of a subroutine subprogram.

R1225 contains-stmt is CONTAINS

R1226 stmt-function-stmt is function-name ([dummy-arg-name-list]) = scalar-expr

Constraint: The scalar-expr may be composed only of constants (literal and named), references to scalar variables and array elements, references to functions and function dummy procedures, and intrinsic operators. If a reference to a statement function appears in scalar-expr, its definition must have been provided earlier in the scoping unit and must not be the name of the statement function being defined.

Constraint: Named constants in scalar-expr must have been declared earlier in the scoping unit or made accessible by use or host association. If array elements appear in scalar-expr, the parent array must have been declared as an array earlier in the scoping unit or made accessible by use or host association. If a scalar variable, array element, function reference, or dummy function reference is typed by the implicit typing rules, its appearance in any subsequent type declaration statement must confirm this implied type and the values of any implied type parameters.

Constraint: The function-name and each dummy-arg-name must be specified, explicitly or implicitly, to be scalar data objects.

Constraint: A given dummy-arg-name may appear only once in any dummy-arg-name-list.

Constraint: Each scalar variable reference in scalar-expr may be either a reference to a dummy argument of the statement function or a reference to a variable local to the same scoping unit as the statement function statement.

A edit descriptor, 113, 129, 173
ABS function, 136, 141, 314
Abstraction, Stepwise Refinement
    and Modules, 34
Accuracy, 65
ACHAR function, 314
ACM (Association for Computing Machinery),
    TOMS (Transactions on Mathematical
Software), 301
ACOS function, 136, 141, 314
Actual argument, 303
Ada, 35
Addition, 60
Addition operator, 59
Additional forms of the DIMENSION
    attribute and DO loop statement, 93
Adjustable array, 303
ADJUSTL function, 141, 314
ADJUSTR function, 314
AIMAG function, 141, 181, 314
AINT function, 141, 314
Algol, 31
Algol 68, 32
Algorithm 48, 298, 303
ALL function, 315
ALLOCATABLE arrays, 102, 228, 279
ALLOCATABLE attribute, 102, 228, 256, 269
ALLOCATE statement, 102, 198, 228, 270, 278
    alternate form, 278
ALLOCATED function, 141, 315
Alphanumeric or character format, A, 173
Alternate RETURN, 292
An index, 84
Analysis, 24
ANINT function, 315
ANY function, 315
APL, 33
Argument, 143, 303
    actual, 222, 303
    array, 227
    assumed length, 228, 303
    assumed shape, 236, 303

assumed size, 293, 303
character, 228
dummy, 222, 304
dummy procedure argument, 279
keyword and optional, 266–267
PRESENT intrinsic function, 267
Argument association, 303
Argument presence, 226
Arithmetic 59–81
Arithmetic assignment statement, 52
Arithmetic IF, 292
Array, 303
Array Constructors, 105, 303
Array Element Ordering, 104, 303
Array Element Ordering and Physical and
    Virtual Memory, 105
Array Functions, 141
Array Sections, 102, 303
Arrays 82–107
and DO loops, 84–85
    allocatable, 102–103, 256, 303
    as parameters, 227
    assumed shape, 236, 303
    assumed size, 293
    automatic, 278, 303
    bounds, 99, 303
    constructor, 105, 303
    conformable, 99
declaration, 84
    deferred shape, 103, 278, 304
    element, 84–85
element ordering, 104, 303
    explicit shape dummy arrays, 227, 235
    extent, 99, 305
index ,84
masked assignment, 106
    rank, 99, 307
    section, 102, 303
    shape, 99, 307
    size, 99, 307
    whole array manipulation, 99–101
Arrays in Fortran, 84

Artificial Language, 21
ASCII character set, 303, 312
ASIN function, 315
ASSIGN and assigned GOTO statements, 293
Assigned FORMAT statements, 293
Assignment, 100
Assignment statement, 52
ASSOCIATED function, 199, 203, 315
Association, 303
    USE association, 264
Assumed length dummy
    argument, 228, 235, 303
Assumed shape arrays, 236, 303
Assumed size array, 293, 303
ATAN function, 316
ATAN2 function, 316
Attribute, 303
Attribute specification, 269
Automatic array, 278, 303

Basic, 34
Bibliography, 8, 14, 18, 26, 42, 80, 151, 169, 194, 209, 239, 258, 290, 302
Binary Representation of Different Integer Kind Type Numbers, 75
Bit Data Type and Representation Model, 71
BIT_SIZE function, 316
Blanks, nulls and zeros, 129
Block IF statement, 153
    ELSE block, 155
    ELSE IF block, 155
Book catalogue, 82
Bottom up, 22
Bounds, 99, 303
BTEST function, 316

C, 34
C++, 39
CALL statement, 222, 270
Case studies, 272–290
    extracting the diagonal elements
        of a matrix, 287

function that returns a variable length
    array, 285–286
generic procedures, 279–286
operator and assignment
    overloading, 286–287
solving ordinary differential
    equations, 272–279
CASE statement, 158, 165, 250–251
    DEFAULT, 159
    syntax, 165
CEILING function, 316
CHAR function, , 177, 316
Character, 171–180
    constant, 303
    functions, 176
    input/ output, 113, 129, 172
    operators, concatenation, 172, 173
    sets, ASCII, 171
    string, 303
    sub-strings, 174
    trailing blanks, 174
CHARACTER statement, 172, 269
Chomsky and Program Language
    Development, 31
CLOSE statement, 213, 270
CMPLX function, 181, 317
Cobol, 30
Coda, 301
Collating sequence, 177, 304
Comments, 51
Common mistakes, 113
Compilation unit, 304
Complete List of Predefined Functions, 140
Complex, 181–184
Complex and Kind Type, 183
Complex conjugate, 182, 317
COMPLEX statement, 181, 269
Complex type, 181
    i/o ,181
Component, 190, 304
Computational Functions, 141
Computer systems, 7–11

Concatenate, 304
Concatentation, 172, 303
Conformable, 99–100, 304
CONJG function, 182, 317
Constant, 304
CONTAINS statement, 149, 249
Contiguous, 304
Continuation character &
Control Structures, 50, 153–170
Conversion functions, 141
Converting from Fortran 77, 292–294
COS function, 135, 317
COSH function, 317
COUNT function, 317
CSHIFT function, 317
CYCLE and EXIT, 162
CYCLE statement, 162, 205–206, 270
Data description statements, 50
Data entity, 304
Data object, 304
Data processing statements, 50
Data structures, 190
Data type, 190, 304
DATE_AND_TIME subroutine, 318
DBLE function, 318
DEALLOCATE statement, 270
Debugging, simple techniques, 300
Declaration, 304
Decremental features, 292
Default kind, 304
Deferred shape arrays, 103, 278, 304
Defined, 304
Defining a subroutine, 222
Derived type definition, 268, 304
Design, 24
Detailed Design, 24
DIGITS function, 319
DIM function, 319
DIMENSION attribute 84, 93, 304
Division, 60
Division operator, 59
DO construct, 97

DO IF EXIT construct, 161
DO loop 85, 87, 94
   construct, 97
   statement, 93,160
   syntax, 166
DO statement 93, 160
DO variable, 86–87
DO WHILE construct, 161
DOT_PRODUCT function, 101, 106, 138, 319
DOUBLE PRECISION, 293
DPROD function, 319
Dummy arguments
   assumed length, 235
   assumed shape, 236, 303
   character, 228, 235
   explicit shape array, 227, 235
   procedures as, 279
Dummy Arguments or Parameters and
   Actual Arguments, 222, 304
Dynamic Data Structures, 195–210
   ASSOCIATED function, 203
   NULLIFY statement, 203
   POINTER attribute, 196
   pointers, 196
   singly linked list, 197
   TARGET attribute, 196
   trees, 200

E edit descriptor, 111
Editors, 16
Efficiency, space time trade off, 299
Elemental, 304
Elemental functions, 137
Elements of a programming language, 49
ELSE block, 155
ELSE IF, 155
ELSEWHERE block, 106
END DO statement, 86
END FILE statement, 270
END FUNCTION statement, 143
END PROGRAM statement, 52
END SELECT statement, 159

Entities, 262, 304
EOSHIFT function, 319
EPSILON function, 320
Errors when reading, 132
Evaluation and testing, 25
Evaluation of arithmetic expressions, 60
Examination marks or results, 83
Exceptional values, 305
EXIT statement, 159, 162, 216, 270
EXP function, 320
Explicit interface, 305
Explicit shape array, 305
Explicit shape dummy arrays, 227, 235
Exponent 71, 72
EXPONENT function, 320
Exponentiation operator, 59
Expression evaluation, 60
Expressions, 100, 305
Extent, 99, 305
External sub-program, 305

F edit descriptor, 110
Feasibility Study and Fact Finding, 24
File manipulation again, 131
File name, 212
File name expression, 212
FILE= specifier, 212
Files, 16, 212–218
    unformatted, 215
Files in Fortran, 212
Fixed fields on input, 126
FLOOR function, 320
FMT=specifier, 115
FORM= specifier, 214
Formal Syntax and Some Additional
    Features, 260–270
FORMAT statement, 109
A edit descriptor, 113, 129, 173
E edit descriptor, 111
F edit descriptor, 110
I edit descriptor, 109
L edit descriptor, 187

X edit descriptor, 112
Formatted data, 214
Formatted i/o, 214
Formatting for a line-printer, 120
Fortran character set, 56
Fortran 90 libraries, 251
FRACTION function, 320
Free format source, 48
Fun, 302
Function reference, 143, 305
Function result, 143, 305
Function syntax, 150
Functions, 135–152, 305, 313–339
    arguments, 143
    array, 141
    computational, 141
    elemental, 137
    generic, 136
    inquiry, 140
    internal, 148
    intrinsic, 135, 313–339
    recursive, 145
    rules and restrictions, 150
    supplying your own, 142
    syntax, 150
    transfer and conversion, 141
    transformational, 137, 307
    user defined, 142

Gaussian elimination, 251–255
GCD, 143
Generating a new line, on both line-printers and
    terminals., 122
Generic, 136, 305
Generic argument, 136
Generic functions, 136
Generic intrinsic function, 136
Generic procedure, 279
Global, 305
Good programming guidelines, 56
GOTO statement, 270, 298–299

Health Warning: Optional Reading, Beginners are Advised to Leave until Later, 67
High Performance Fortran, HPF, 41
Higher dimension arrays, 91
History of Operating Systems, 12
    The 1940s, 12
    The 1950s, 12
    The 1960s, 12
    The 1960s and 1970s, 12
    The 1980s, 13
Hoare's Quick Sort algorithm, 229
Host association, 305
HPF, 41
HUGE function, 320

I edit descriptor, 109
I/O, 50, 187
    foolproof, 215
IACHAR function, 321
IAND function, 321
IBCLR function, 321
IBITS function, 321
IBSET function, 321
ICHAR function, 321
ICON, 37
IEEE, 72
IEOR function, 322
IF statement
    block, 153, 185
    ELSE block, 155
    ELSE IF block, 155
    syntax, 166
Implicit interface, 305
Implied DO loops, 119
IMSL, 301
Index, 84
INDEX function, 176, 322
Inner Product of two Sparse Vectors, 204
INQUIRE statement, 270
Inquiry Functions, 140, 305
INT function, 181

Integer Data Type and Representation Model, 71
INTEGER type specification, 53–55, 269
Integers, I format, 109, 126
I/O, 109–110, 126
Intent, 223
INTENT attribute, 223–224, 235
Interface, 222
Interface block, 222–223, 234, 269, 305
    mandatory, 234
Interface body, 305
Internal functions, 148
Internal procedure, 235, 260, 268, 305
Internal subroutine, 235
    and scope, 235
Intrinsic functions, 135, 305
Intrinsic procedures, 305
I/O (Input/ Output), 108–133
    Character, 113, 129, 172
    Complex, 181
    Integer, 109–110, 126
    Logical, 187–188
    Real, 110–111, 126–129
IOR function, 141
ISHFT function, 322
ISHFTC function, 323
ISO, i

Keyword and optional arguments, 226
Kind, 67, 305
KIND function 67, 323
Kind type parameters, 68, 256, 306

L edit descriptor, 187
Language extension, 306
LBOUND function, 323
LEN function, 176, 323
LEN_TRIM function, 324
LGE function, 177, 324
LGT function, 177, 324
Linked list, 197
Linker, 306

LINPACK, 252, 258
Lisp, 32
LLE function, 177, 324
LLT function, 177, 324
Local entity, 306
Local variables, 144, 224
Local variables and the SAVE attribute, 224
LOG function, 324
LOG10 function, 324
Logical 184–194
   data type, 184
   expression, 154, 185
   I/O, 187–188
   operators, .AND., .NOT., .OR., 185–186
   statement, 185
   variable, 185
LOGICAL function, 325
Logical operators, 185
Logical type, 185
LOGICAL type specification, 185, 269
Logical variable, 185
Logo, 36

Main program, 267, 306
Maintenance, 25
Mantissa 71, 72
Masked array assignment, 106
Masked Array Assignment and the
   WHERE Statement., 106
MATMUL function, 101, 325
MAX function, 325
MAXEPONENT function, 325
MAXLOC function, 256, 325
MAXVAL function, , 256, 326
MERGE function, 326
MIN function, 327
MINEXPONENT function, 327
MINLOC function, 327
MINVAL function, 327
Miscellanea, 296–301
MOD function, 328
Modula, 35

Modula 2, 36
Module, 260, 268, 306
Modules, 242–258, 260, 268, 296
Module procedure, 249, 306
Modules
   and packaging, 288
   and scope, 264
   containing procedures, 249
   defining kind type parameters, 256
   for derived data types, 245
   for explicit procedures interfaces, 248
   for global data, 242
   PUBLIC and PRIVATE attributes, 264–265
   usage and compilation, 256
   USE, ONLY and rename, 265–266
MODULO function, 328
Monthly Rainfall, 83, 85
More foolproof i/o, 215
Multi-user systems, 17
Multiple statements, 55
Multiplication operator, 59
MVBITS function, 329

NAG, 301
NAGWare f90 tools, 300
Name, type and value, 52, 306
Name association, 306
Natural Language, 20
NEAREST function, 329
Nesting, 306
Netlib, 301
Network Sources, 41
Networked Systems, 16
Networking, 13
NINT function, 329
Nonexecutable statement, 306
NOT function, 329
Notations, 21
NULLIFY statement, 203, 270
Numerical Software Sources, 300

Oberon and Oberon 2, 38

Object Orientated Programming, 38
Object file, 306
Obsolescent Features, 292
OPEN (and CLOSE), 114, 131, 212–213
OPEN statement, 114, 212–213, 270
    parameters, 213–214
Operating systems, 12–15
Operator and assignment overloading, 286
Operators, 59
Operator hierarchy, 186
Optional arguments, 266
Ordinary differential equations, first order, 272
Output, 109–125
Overview, 2

PACK function, 287, 329
Parameter, 222, 306
Parameter attribute 64, 65
Pascal, 33
PAUSE Statement, 293
Peoples Weight's, 87
PL/1 and Algol 68, 32
Pointer, 196, 306
Pointer association, 196, 306
POINTER attribute, 196
Positional Number Systems, 70
Postscript, TeX and LaTeX, 37
Pre-Defined Subroutines, 142
Precision, 65, 306
Precision and size of numbers, 65
PRECISION function, 330
PRESENT function, 330
Pretty print, 300
PRINT statement, 109, 270
PRIVATE statement, 265
Problem Definition, 24
Problem solving, 20–28
Procedure – Function or Subroutine, 260
    as argument, 279
    ending, 268
    generic, 279
    heading, 268
    interface, 306
    internal, 268
PRODUCT function, 330
Program, 260, 306
    unit, 306
Program Development and Software Engineering, 296
Program Language Development and Engineering, 29
Program testing, 299
Programming languages, 29–47
    Ada, 35, 309
    Algol, 31, 309
    Algol 68, 32
    APL, 33, 309
    Basic, 34, 309
    C, 34, 309–310
    C++, 39, 310
    Cobol, 30, 310
    Fortran IV, 30
    Fortran 66, 30
    Fortran 90, 40, 310
    ICON, 37, 310
    LaTeX, 36
    Lisp, 32, 310
    Logo, 36, 310
    Modula, 35
    Modula 2, 36, 310–311
    Oberon, 38, 311
    Oberon 2, 38, 311
    Pascal, 33, 311
    PL/1, 32
    Postscript, 37, 311
    Prolog, 37, 311
    Simula, 33, 311
    Smalltalk, 39, 311
    Snobol, 311
    SQL, 37
    TeX, 37
Programming style, 302
Programming Style – Programs should be Easy to Read, 297

Programming Style – Programs should
    Behave Well, 297
Prolog, 37
PUBLIC and PRIVATE attribute, 264–265

Quick sort algorithm, 229

RADIX function, 331
RANDOM_NUMBER function, 331
RANDOM_SEED function, 331
Range and precision, 67
RANGE function, 332
Rank, 99, 307
READ statement, 130, 270
Reading, 130
Reading in Data, 126–134
Real and double precision DO Control
    Variables, 292
Real Data Type and Representation Model, 72
REAL function, 181, 332
REAL type specification, 53–55, 269
I/O 110–111, 126–129
Reals, E format, 111
Reals, F format, 110
Reals, the F and E formats, 126
Record, 117
Recursion, 298, 307
Recursion and Problem Solving, 151
Recursive functions, 145
Recursive subroutines, 235
Reference, 307
Referencing a subroutine, 222
Relational expression, 307
Relational operator, 155
Rename, 265
REPEAT function, 141
REPEAT UNTIL loop, 161
Repetition, 116
RESHAPE function, 105, 332
RETURN statement, 270
REWIND, 270
Rounding and truncation, 61

RRSPACING function, 333
Runge Kutta Merson algorithm, 272
Rules and Restrictions, 150

SAVE attribute, 224
Scalar, 307
Scalar variable, 307
SCALE function, 333
SCAN function, 333
Scope, 144, 224, 307
Scope and association, 262–263
Scope of variables, 144, 224
Scoping unit, 144, 307
Second Generation Languages, 32
SELECT CASE statement, 158–160
SELECTED_INT_KIND function 69, 333
SELECTED_REAL_KIND function 69, 334
Selecting different INTEGER Kinds, 69
Selecting different REAL Kinds, 69
SET_EXPONENT function, 334
Shape, 307
Shape conformance, 307
SHAPE function, 334
Shared DO termination and
    non ENDDO termination, 292
SIGN function, 334
Simple Debugging Techniques, 300
Simple Subroutine Example, 219
Simula, 33
SIN function, 135, 334
Singly linked list, 197
SINH function, 335
Size, 99, 307
SIZE function, 335
Skipping spaces and lines, 130
Smalltalk, 39
Snobol, 32
Software, 8
Some Other Strands in Language
    Development, 34
Source file, 307
Source form, 48, 55

SPACING function, 335
Sparse matrix problems, 203
Specification construct, 268
Specification statement, 269
Specifying Kind Types for Literal Integer and Real Constants, 70
SPREAD function, 335
SQL, 37
SQRT function, 336
Stand Alone Systems, 16
Standardisation, 35
Statement, 307
    ordering, 261
Statement separator, (;), 48
Status of the Action Carried out in the Subroutine, 224
Stepwise refinement, 23
STOP statement, 270
Stride, 307
Structure, 307
Structured programming, 34
Structured programming and the GOTO statement, 298
Subprogram, 267, 307
Subroutine, 219–241, 307
    actual arguments, 222
    assumed shape arrays, 236
    automatic arrays as parameters, 278
    dummy arguments, 222
    dummy procedure arguments, 279
    external, 305
    interface blocks, 222–223, 234
    internal, 235
    internal and scope, 235
    INTENT attribute, 223–224, 235
    introduction to arrays as parameters, 227
    keyword and optional arguments, 266–267
    local variables, 224
        SAVE attribute, 224
    rank 2 and higher arrays as parameters, 236
    recursive, 235
Subscript, 86, 307

Substring, 307
Subtraction operator, 59
SUM function, 336
Summary of how to select the appropriate KIND type, 77
Summary of options on OPEN, 214
Supplying your own functions, 142
Syntax, 49
Syntax summary of some frequently used fortran constructs, 267
Systems Analysis and Design, 23
    analysis, 24
    design, 24
    detailed design, 24
    evaluation and testing, 25
    feasibility study, 24
    implementation, 25
    maintenance, 25
    problem definition, 24
SYSTEM_CLOCK function, 230–234, 336

Tables of data, 82
TAN function, 135, 136, 141, 337
TANH function, 141, 337
Target, 203, 307
TARGET Attribute, 196
Telephone directory, 82
Terminology, 99
TINY function, 337
Top down, 22
Transfer and Conversion Functions, 141
TRANSFER function, 337
Transformational Functions, 137, 307
TRANSPOSE function, 337
Trees, 200
TRIM function, 338
Truncation, 61, 307
Truth Tables, 186
Type, name and value, 52
Type declaration, 307
Type definition, 190
Typing, 52, 307

UBOUND function, 141
Underflow, 308
Unformatted files, 214–215
UNPACK function, 338
Upper bound, 224
USE association, 264
USE, 242, 269
USE, ONLY and rename, 265–266
User defined types, 189–194, 308
    component, 190, 304
    nested user defined types, 192
    records, 190
    type definition, 190
Using Linked Lists for Sparse Matrix
    Problems, 203
Using computer systems, 16–19

Value, 52
Variable Definition, 191
Variables, 52, 308
Variables, name type and value, 52
Vector, 81
VERIFY function, 339
Von Neumann, 29

What is a programming language ?, 29
WHERE statement, 106
    syntax, 166, 270
WHILE loop, 161
Whole array manipulation, 99
Why Bother?, 225
Workspace, 278
WRITE statement, 115, 270
Writing, 115

X edit descriptor, 112

3834

AUG 1 8 1998

ENGINEERING LIBRARY